Advanced Maths Essentials
Core 1 for AQA

Welcome to Advanced Maths Essentials: Core 1 for AQA. This book will help you to improve your examination performance by focusing on all the essential maths skills you will need in your AQA Core 1 examination.

The book contains scores of worked examples, each with clearly set-out steps to help solve the problem. You can then apply the steps to solve the Skills Check questions in the book and past exam questions at the end of each chapter. If you feel you need extra practice on any topic, you can try the Skills Check Extra exercises on the accompanying CD-ROM. At the back of this book there is a sample exam-style paper to help you test yourself before the big day.

Some of the questions in the book have a  symbol next to them. These questions have a PowerPoint® solution (on the CD-ROM) that guides you through suggested steps in solving the problem and setting out your answer clearly.

Using the CD-ROM

To use the accompanying CD-ROM simply put the disc in your CD-ROM drive, and the menu should appear automatically. If it doesn't automatically run on your PC:

1. Select the My Computer icon on your desktop.
2. Select the CD-ROM drive icon.
3. Select Open.
4. Select core1_for _aqa.exe.

If you don't have PowerPoint® on your computer you can download PowerPoint 2003 Viewer®. This will allow you to view and print the presentations. Download the viewer from http://www.microsoft.com

Pearson Education Limited
Edinburgh Gate
Harlow
Essex
CM20 2JE
England
www.longman.co.uk

First published 2004
ISBN   0 582 83677 8

Design by Ken Vail Graphic Design

Cover design by Raven Design

Typeset by Tech-Set, Gateshead

Printed in the U.K. by Scotprint, Haddington

The publisher's policy is to use paper manufactured from sustainable forests.

The Publisher wishes to draw attention to the Single-User Licence Agreement situated at the back of the book. Please read this agreement carefully before installing and using the CD-ROM.

We are grateful for permission from the Assessment and Qualifications Alliance to reproduce past exam questions. All such questions have a reference in the margin. The Assessment and Qualifications Alliance can accept no responsibility whatsoever for accuracy of any solutions or answers to these questions.

Every effort has been made to ensure that the structure and level of sample question papers matches the current specification requirements and that solutions are accurate. However, the publisher can accept no responsibility whatsoever for accuracy of any solutions or answers to these questions. Any such solutions or answers may not necessarily constitute all possible solutions.

1 Algebra

1.1 Surds

Use and manipulation of surds.

Some square roots cannot be written as a fraction or as a terminating or recurring decimal. They are irrational and are called **surds**.

Examples of surds are $\sqrt{2}$, $\sqrt{3}$, $\sqrt{5}$.

Note:
The value given on the calculator for a surd is only an approximation.

Like surds can be combined by **adding** or **subtracting**:

$$5\sqrt{3} + 4\sqrt{3} + 6\sqrt{5} - 2\sqrt{5} = 9\sqrt{3} + 4\sqrt{5}$$

When **multiplying**, remember that $\sqrt{p} \times \sqrt{q} = \sqrt{p \times q}$.

$$\sqrt{2} \times \sqrt{3} = \sqrt{2 \times 3} = \sqrt{6}$$

When **squaring**, use the special case $\sqrt{p} \times \sqrt{p} = \sqrt{p^2} = p$.

$$(3\sqrt{5})^2 = 3\sqrt{5} \times 3\sqrt{5} = 9 \times \sqrt{5 \times 5} = 9 \times 5 = 45$$

To **simplify** a surd such as $\sqrt{12}$ or $\sqrt{180}$, write it in the form $k\sqrt{a}$, where a does not have any factors that are square numbers.

$$\sqrt{12} = \sqrt{2 \times 2 \times 3} = \sqrt{2^2 \times 3} = \sqrt{2^2} \times \sqrt{3} = 2 \times \sqrt{3} = 2\sqrt{3}$$

$$\sqrt{180} = \sqrt{2 \times 2 \times 3 \times 3 \times 5} = \sqrt{2^2 \times 3^2 \times 5} = 2 \times 3 \times \sqrt{5} = 6\sqrt{5}$$

Tip:
Split the number into its prime factors to help spot factors that are square numbers.

Example 1.1   **a**   Write $\sqrt{8}$ in the form $k\sqrt{2}$, where k is an integer.

b   Hence simplify $6\sqrt{2} + 5\sqrt{8}$.

Step 1: Simplify the surd.   **a** $\sqrt{8} = \sqrt{4 \times 2} = \sqrt{4} \times \sqrt{2} = 2\sqrt{2}$

Step 2: Add like surds.   **b** $6\sqrt{2} + 5\sqrt{8} = 6\sqrt{2} + 5 \times 2\sqrt{2}$
$$= 6\sqrt{2} + 10\sqrt{2}$$
$$= 16\sqrt{2}$$

Example 1.2   Expand and simplify $(2 - \sqrt{3})(5 + \sqrt{3})$.

Step 1: Expand the brackets. $(2 - \sqrt{3})(5 + \sqrt{3}) = 10 + 2\sqrt{3} - 5\sqrt{3} - (\sqrt{3})^2$
Step 2: Simplify each term. $= 10 - 3\sqrt{3} - 3$
Step 3: Add like terms. $= 7 - 3\sqrt{3}$

Tip:
$(\sqrt{3})^2 = 3$

If you recognise the **difference of two squares** you can multiply quickly, using $(a + b)(a - b) = a^2 - b^2$, for example

$$(3 + \sqrt{2})(3 - \sqrt{2}) = 3^2 - (\sqrt{2})^2 = 9 - 2 = 7$$

Note:
This product is rational.

To simplify fractions with a surd in the denominator, use a special technique called **rationalising the denominator**. This involves forming an equivalent fraction by multiplying both the numerator and denominator by a quantity that makes the denominator rational.

Example 1.3 Express $\dfrac{12}{\sqrt{3}}$ in the form $k\sqrt{3}$, where k is an integer.

Step 1: Rationalise the denominator.

$$\frac{12}{\sqrt{3}} = \frac{12 \times \sqrt{3}}{\sqrt{3} \times \sqrt{3}}$$

Step 2: Simplify if possible.

$$= \frac{12\sqrt{3}}{3} = 4\sqrt{3}$$

Example 1.4 Simplify $\dfrac{2\sqrt{3} + 1}{\sqrt{3} - 1}$.

Step 1: Rationalise the denominator and simplify.

$$\frac{2\sqrt{3} + 1}{\sqrt{3} - 1} = \frac{(2\sqrt{3} + 1)(\sqrt{3} + 1)}{(\sqrt{3} - 1)(\sqrt{3} + 1)}$$

$$= \frac{6 + 3\sqrt{3} + 1}{3 - 1^2}$$

$$= \frac{7 + 3\sqrt{3}}{2}$$

Example 1.5 A rectangle $ABCD$ has area $20\sqrt{2}$ cm^2 and length $AB = 4\sqrt{5}$ cm.

Giving your answers in simplified surd form, find **a** the length of BC

b the length of DB.

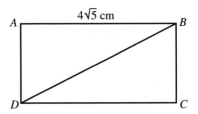

Step 1: Use the formula for the area of a rectangle to find the length.

a $\quad$ Area $= AB \times BC$

$\Rightarrow 20\sqrt{2} = 4\sqrt{5} \times BC$

$$BC = \frac{20\sqrt{2}}{4\sqrt{5}}$$

$$= \frac{20\sqrt{2} \times \sqrt{5}}{4\sqrt{5} \times \sqrt{5}}$$

$$= \frac{20\sqrt{10}}{4 \times 5}$$

$$= \sqrt{10}$$

Length $BC = \sqrt{10}$ cm

Step 2: Use Pythagoras' Theorem in triangle ADB.

b By Pythagoras' theorem

$$DB^2 = AD^2 + AB^2$$
$$= (\sqrt{10})^2 + (4\sqrt{5})^2$$
$$= 10 + 16\,(5)$$
$$= 90$$
$$DB = \sqrt{90}$$
$$= \sqrt{9 \times 10}$$
$$= 3\sqrt{10}$$

Length $DB = 3\sqrt{10}$ cm

For further examples using surds, see Sections 1.6 (Quadratic equations) and 1.8 (Inequalities).

1 Simplify **a** $\sqrt{50}$ **b** $\sqrt{32}$ **c** $\sqrt{98}$ **d** $\sqrt{12} + 5\sqrt{3}$ **e** $3\sqrt{7} - \sqrt{5} + 4\sqrt{7} - 3\sqrt{5}$.

2 Simplify the following, where $p = \sqrt{3}$, $q = \sqrt{2}$, $r = \sqrt{12}$, $s = \sqrt{18}$.

    a $3r + 5p$ **b** $s - q$ **c** $(p - q)^2$ **d** $(5q)^2$

    e q^3 **f** $\dfrac{s}{q}$ **g** $\dfrac{1}{r - p}$ **h** $\dfrac{q}{q + p}$

3 Simplify   **a** $(\sqrt{11} + 1)^2$ **b** $(\sqrt{11} + 1)(\sqrt{11} - 1)$ **c** $\dfrac{\sqrt{11} + 1}{\sqrt{11} - 1}$.

 4 Triangle ABC is right-angled at B, $AB = \dfrac{3}{\sqrt{5}}$ cm, $BC = 5\sqrt{2}$ cm.

    Its area is $p\sqrt{10}$ cm², where p is rational. Find the value of p.

5 **a** Express $\sqrt{48}$ and $\dfrac{6}{\sqrt{3}}$ in the form $k\sqrt{3}$, where k is an integer.

    b Hence write $\sqrt{48} + \dfrac{6}{\sqrt{3}}$ in the form $p\sqrt{3}$, where p is an integer.

 6 **a** Express $(\sqrt{7} + 1)^2$ in the form $a + b\sqrt{7}$, where a and b are integers.

    b Express $\dfrac{\sqrt{7} + 1}{\sqrt{7} - 1}$ in the form $c + d\sqrt{7}$, where c and d are rational numbers.

7 **a** Given that $(3 + \sqrt{5})(4 - \sqrt{5}) = p + q\sqrt{5}$, where p and q are integers, find p and q.

    b Given that $\dfrac{3 + \sqrt{5}}{4 + \sqrt{5}} = r + s\sqrt{5}$, where r and s are rational numbers, find r and s.

SKILLS CHECK **1A EXTRA** is on the CD

1.2 Quadratic functions

Quadratic functions and their graphs.

The function $f(x) = ax^2 + bx + c$, where $a \neq 0$, is a **quadratic** function.

> **Note:**
> If a is zero, there is no x^2 term.

The graph of $y = ax^2 + bx + c$ is called a **parabola**. It is a symmetrical curve with one turning point, called the **vertex**.

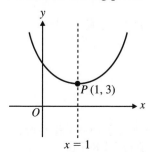

If $a > 0$, the turning point is a **minimum** turning point.

If $a < 0$, the turning point is a **maximum** turning point.

The diagram shows $y = 2x^2 - 4x + 5$.

There is a minimum turning point at the vertex $P(1, 3)$.

The minimum value of y is 3.

The axis of symmetry is the line $x = 1$.

> **Note:**
> See Section 1.13 for more on finding the turning point and drawing parabolas.

> **Note:**
> See Section 3.6 for calculus methods.

3

1.3 The discriminant

The discriminant of a quadratic function.

Given a quadratic function, $f(x) = ax^2 + bx + c$, the **discriminant** is the expression $b^2 - 4ac$.

It can be used to find

- the number of solutions (or roots) of the quadratic equation $ax^2 + bx + c = 0$ (Section 1.6)

- whether a quadratic curve crosses the x-axis at two points, touches it at one point or does not meet or cross the x-axis (Section 1.13)

- whether a line and a curve, or two curves, intersect in two places, meet at a point or do not intersect (Section 1.14).

1.4 Factorising quadratic polynomials

Factorisation of quadratic polynomials.

When you multiply two linear expressions in x, you get a quadratic polynomial, for example $(3x - 2)(4x + 5) = 12x^2 + 7x - 10$.

To **factorise** a quadratic polynomial, reverse the process and express it as the product of two linear functions.

Some factorising can be done by guesswork, especially when the coefficient of x^2 is 1, for example:

$$x^2 + 8x = x(x + 8) \qquad \text{(common factor)}$$
$$x^2 - 9 = (x - 3)(x + 3) \quad \text{(difference of two squares)}$$
$$x^2 + 8x + 7 = (x + 1)(x + 7)$$
$$x^2 + 8x - 9 = (x - 1)(x + 9)$$

Note:
Not all quadratic expressions can be factorised.

Tip:
Always look for a common factor first.

When guesswork becomes time-consuming, it may be quicker to use a method, as in the following example.

To factorise $12x^2 + 7x - 10$:

- Compare with $ax^2 + bx + c$: $a = 12, b = 7, c = -10$

- Calculate $a \times c$: $ac = -120$

- Find factors of ac that add to give b:

Factors of -120	Sum
120×-1	119
60×-2	58
20×-6	14
15×-8	7

The factors of -120 that add to give 7 are 15 and -8.

Tip:
Since ac is negative and b is positive, the signs must be different and the 'larger' number must be positive.

Tip:
Keep practising and you will find that your guesses for the two numbers will become more efficient.

- Write $7x$ as $15x - 8x$ (or as $-8x + 15x$):
$$12x^2 + 7x - 10 = 12x^2 + 15x - 8x - 10$$

- Group the terms in pairs, taking care with signs:
$$12x^2 + 7x - 10 = (12x^2 + 15x) - (8x + 10)$$

Tip:
Remember, if you put a bracket immediately after a minus, you must change the sign in the bracket.

- Take out any common factors from each of the pairs:
$$12x^2 + 7x - 10 = 3x(4x + 5) - 2(4x + 5)$$

- Finally take out a common expression:
$$12x^2 + 7x - 10 = (4x + 5)(3x - 2)$$

This method also works when the coefficient of x^2 is negative. However, to avoid problems with signs, it may be better to take out a factor of -1 first, as in Example 1.6.

Example 1.6 Factorise $f(x) = -9x^2 + 18x - 5$.

Step 1: Write $f(x)$ in the form $-g(x)$.

$$-9x^2 + 18x - 5 = -(9x^2 - 18x + 5)$$

Tip:
Take out a factor of -1.

Now consider $9x^2 - 18x + 5$.

Step 2: Identify a, b and c.

$a = 9$, $b = -18$, $c = 5$

Step 3: Find factors of ac that add up to give b.

$ac = 9 \times 5 = 45$

Factors of 45 that add to give -18 are -15 and -3.

Step 4: Rewrite bx using the factors found in Step 2.

Step 5: Factorise the first two and last two terms. Then take out the common bracket.

Step 6: Multiply *one* of the brackets by -1.

$$\begin{aligned}
-9x^2 + 18x - 5 &= -(9x^2 - 18x + 5) \\
&= -(9x^2 - 15x - 3x + 5) \\
&= -((9x^2 - 15x) - (3x - 5)) \\
&= -(3x(3x - 5) - 1(3x - 5)) \\
&= -(3x - 5)(3x - 1) \\
&= (3x - 5)(1 - 3x)
\end{aligned}$$

Tip:
It does not matter which bracket you multiply by -1, but do not multiply both. In this example
$$-(3x - 1) = -3x + 1$$
$$= 1 - 3x.$$

Important note:
In the special case when $a = 1$, the quadratic expression can be factorised very easily. In this case, ac is the same as c, so you just find two numbers that multiply to give c, the constant term, and add to give b, the x term. These numbers then go in the brackets.

Consider $x^2 - x - 6$:

Factors of -6 that add to give -1 are -3 and 2.

So $x^2 - x - 6 = (x - 3)(x + 2)$.

1.5 Completing the square

Completing the square.

Remember the pattern when squaring linear functions:
$$(x + p)^2 = x^2 + 2px + p^2 \qquad (x - p)^2 = x^2 - 2px + p^2$$

Recall:
$(x + 5)^2 = x^2 + 10x + 25$
$(x - 3)^2 = x^2 - 6x + 9$

Now rearrange these:
$$x^2 + 2px = (x + p)^2 - p^2 \qquad x^2 - 2px = (x - p)^2 - p^2$$

Tip:
The number in the bracket is half the coefficient of x. Then subtract the square of this number.

For example:
$$\begin{aligned}
x^2 + 10x &= (x + 5)^2 - 25 & (p = 5) \\
x^2 - 6x &= (x - 3)^2 - 9 & (p = 3) \\
x^2 + 3x &= (x + \tfrac{3}{2})^2 - \tfrac{9}{4} & (p = \tfrac{3}{2})
\end{aligned}$$

This process is called **completing the square**. It can also be applied to quadratic expressions with a constant term, for example:
$$\begin{aligned}
x^2 + 10x + 30 &= (x + 5)^2 - 25 + 30 = (x + 5)^2 + 5 \\
x^2 + 10x - 20 &= (x + 5)^2 - 25 - 20 = (x + 5)^2 - 45
\end{aligned}$$

Note:
Use this technique to find the coordinates of the centre of a circle. See Section 2.3.

When $a \neq 1$:

$$3x^2 + 2x + 1 = 3(x^2 + \tfrac{2}{3}x) + 1$$
$$= 3((x + \tfrac{1}{3})^2 - \tfrac{1}{9}) + 1$$
$$= 3(x + \tfrac{1}{3})^2 - \tfrac{1}{3} + 1$$
$$= 3(x + \tfrac{1}{3})^2 + \tfrac{2}{3}$$

Tip:
Take out a factor of 3 from the x^2 and x terms first.

When $a < 0$:

$$1 - 10x - x^2 = -(x^2 + 10x - 1)$$
$$= -(((x + 5)^2 - 25) - 1)$$
$$= -((x + 5)^2 - 26)$$
$$= 26 - (x + 5)^2$$

Tip:
Take out a factor of -1 first. Remember to multiply through by it at the end.

Completing the square involves writing the quadratic expression $ax^2 + bx + c$ in the form $A(x + B)^2 + C$. You may prefer to complete the square using this identity, as in Example 1.7.

Example 1.7 Write $3x^2 + 2x + 1$ in the form $A(x + B)^2 + C$.

Step 1: Expand $A(x + B)^2 + C$.

$$A(x + B)^2 + C \equiv A(x + B)(x + B) + C$$
$$\equiv A(x^2 + 2Bx + B^2) + C$$
$$\equiv Ax^2 + 2ABx + AB^2 + C$$

Note:
The $\equiv$ symbol indicates that the expressions are identically equal, for all values of x.

Step 2: Compare coefficients with the original expression.

Step 3: Evaluate A, B and C.

So $3x^2 + 2x + 1 \equiv Ax^2 + 2ABx + AB^2 + C$

Coefficient of x^2: $\quad 3 = A$

Coefficient of x: $\quad 2 = 2AB \Rightarrow B = \tfrac{1}{3}$

Constant term: $\quad 1 = AB^2 + C \Rightarrow C = 1 - AB^2 = 1 - 3(\tfrac{1}{3})^2 = \tfrac{2}{3}$

Substituting $A = 3$, $B = \tfrac{1}{3}$ and $C = \tfrac{2}{3}$ gives

$$3x^2 + 2x + 1 \equiv 3(x + \tfrac{1}{3})^2 + \tfrac{2}{3}$$

Applications of completing the square

When a quadratic function f(x) is in completed square form it is easy to find its maximum or minimum value and also the turning point and axis of symmetry of the curve $y =$ f(x).

Example 1.8 It is given that f$(x) = 2x^2 - 4x + 5$.

a Write f(x) in the form $A(x + B)^2 + C$.

b Write down the least value of f(x) and state the value of x at which this occurs.

c Hence state the coordinates of the minimum turning point on the curve $y =$ f(x).

d Write down the equation of the axis of symmetry of the curve.

Step 1: Expand $A(x + B)^2 + C$.

Step 2: Compare coefficients with the original expression, and evaluate A, B and C.

a $2x^2 - 4x + 5 \equiv A(x + B)^2 + C$
$$\equiv Ax^2 + 2ABx + AB^2 + C$$

Coefficient of x^2: $\quad 2 = A$

Coefficient of x: $\quad -4 = 2AB \Rightarrow B = -1$

Constant term: $\quad 5 = AB^2 + C$
$$\Rightarrow C = 5 - AB^2 = 5 - 2(-1)^2 = 3$$

So f$(x) = 2x^2 - 4x + 5 = 2(x - 1)^2 + 3$

Step 3: Find value of x such that $A(x + B)^2 = 0$.

Step 4: Substitute into f(x).

b When $x = 1$, $2(x - 1)^2 = 0$,
so f$(x) = 0 + 3 = 3$.
For *all other values* of x,
$2(x - 1)^2 > 0 \Rightarrow$ f$(x) > 3$.
Since f$(x) \geqslant 3$, the least value of f(x) is 3 and it occurs when $x = 1$.

Tip:
Focus on the squared part of the expression.

Step 5: State $(-B, C)$.

c The curve $y = 2x^2 - 4x + 5$ has a minimum turning point at $(1, 3)$.

Note:
See Section 1.2 for the sketch of this curve.

Step 6: State the equation of the axis of symmetry.

d The axis of symmetry goes through the turning point so its equation is $x = 1$.

In general, f$(x) = A(x + B)^2 + C$ has a turning point at $(-B, C)$.
This is a minimum if $A > 0$ and a maximum if $A < 0$.

Tip:
See Translations (Section 1.15).

Note:
You could use calculus to find the turning point and the maximum or minimum value (see Section 3.6).

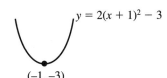

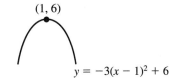

Tip:
If you write $y = -3(x - 1)^2 + 6$ as $y = 6 - 3(x - 1)^2$ it is easier to see that the maximum value is 6.

SKILLS CHECK 1B: Quadratic functions – factorising and completing the square

1 Factorise these expressions.

a $x^2 + 5x$ **b** $x^2 - 2x + 1$ **c** $a^2 - 16$

d $x^2 - 5x - 6$ **e** $x^2 + 13x - 30$ **f** $2x^2 - 8x$

2 Factorise these expressions.

 a $2x^2 + 7x + 6$ **b** $5x^2 - 14x - 3$ **c** $x^3 - 4x^2 - 21x$

d $3y^2 + 4y - 4$ **e** $12 - 4x - 40x^2$ **f** $4x^2 - 25$

3 Write each of these quadratic expressions in the form $A(x + B)^2 + C$.

a $x^2 + 6x + 8$ **b** $x^2 - 12x - 3$ **c** $x^2 + 5x - 2$

4 Write the quadratic expressions in Question **2** in the form $A(x + B)^2 + C$.

5 It is given that $x^2 - 4x + 7 = (x - p)^2 + q$.

a Find p and q.

b Hence write down the coordinates of the vertex of the curve $y = x^2 - 4x + 8$.

c State whether the vertex is a maximum or a minimum turning point.

d Write down the equation of the axis of symmetry of the curve.

 6 It is given that f$(x) = 4x^2 + 8x + 1$.

a Write f(x) in the form $A(x + B)^2 + C$.

b Find the coordinates of the vertex of the curve $y = $ f(x), stating whether it is a maximum or a minimum turning point.

c Write down the equation of the axis of symmetry of the curve.

7 **a** Write f$(x) = 14 - 4x - x^2$ in the form $C - (x + B)^2$.

b Hence state the maximum value of f(x).

c Write down the coordinates of the vertex of the curve $y = 14 - 4x - x^2$.

SKILLS CHECK 1B EXTRA is on the CD

A quadratic equation has the form $ax^2 + bx + c = 0$, where $a \neq 0$. It can have two real roots or one real root or no real roots.

There are several ways of solving quadratic equations. Here are three important techniques.

Solving by factorising

If the quadratic expression can be factorised into the product of two linear functions, solve the quadratic equation using the fact that if $p \times q = 0$, then either $p = 0$ or $q = 0$.

Example 1.9 **a** Factorise $f(x) = 6x^2 + 11x + 3$.

b Hence solve $6x^2 + 11x + 3 = 0$.

Step 1: Factorise the quadratic expression.

a $f(x) = 6x^2 + 11x + 3$
$ = 6x^2 + 9x + 2x + 3$
$ = (6x^2 + 9x) + (2x + 3)$
$ = 3x(2x + 3) + 1(2x + 3)$
$ = (2x + 3)(3x + 1)$

Working for factorisation:
$a = 6, b = 11, c = 3$
$ac = 18$
Factors of 18 that add to give 11 are 9 and 2.

b $ 6x^2 + 11x + 3 = 0$

Step 2: Use $pq = 0 \Rightarrow$ $p = 0$ or $q = 0$ and solve the resulting linear equations.

$\Rightarrow (2x + 3)(3x + 1) = 0$
$\Rightarrow 2x + 3 = 0 \qquad$ or $\qquad 3x + 1 = 0$
$ x = -\frac{3}{2} \qquad\qquad\qquad x = -\frac{1}{3}$

Example 1.10 Solve $8x^2 + 4x - 2 = -x(2x - 3)$.

Step 1: Rearrange the equation to $f(x) = 0$.

$8x^2 + 4x - 2 = -2x^2 + 3x$
$10x^2 + x - 2 = 0$

Step 2: Factorise the quadratic expression.

$(2x + 1)(5x - 2) = 0$

Step 3: Use $pq = 0 \Rightarrow$ $p = 0$ or $q = 0$ and solve the resulting linear equations.

$\Rightarrow 2x + 1 = 0 \qquad$ or $\qquad 5x - 2 = 0$
$ x = -\frac{1}{2} \qquad\qquad\qquad x = \frac{2}{5}$

Solving by completing the square

Complete the square for the quadratic expression and then rearrange to make x the subject.

Example 1.11 **a** Write $x^2 - 8x + 9$ in the form $(x - p)^2 - q$.

b Hence solve the equation $x^2 - 8x + 9 = 0$, leaving your answers in surd form.

Step 1: Complete the square

a $x^2 - 8x + 9 = (x - 4)^2 - 16 + 9$
$ = (x - 4)^2 - 7$

b
$$x^2 - 8x + 9 = 0$$
$$\Rightarrow (x-4)^2 - 7 = 0$$

Tip:
Bring x to one side and all other terms to the other side.

Step 2: Make x the subject.

(Add 7) $\qquad\qquad (x-4)^2 = 7$

(Square root both sides) $\quad x - 4 = \pm\sqrt{7}$

(Add 4) $\qquad\qquad\qquad x = 4 \pm\sqrt{7}$

The two solutions are $x = 4 + \sqrt{7}$ and $x = 4 - \sqrt{7}$.

Tip:
Remember to include $\pm$ when taking the square root. This gives two distinct roots.

Example 1.12

a Write $6x^2 + 11x + 3$ in the form $A(x + B)^2 + C$.

b Hence solve $6x^2 + 11x + 3 = 0$.

Note:
This equation was solved by factorising in Example 1.9.

Step 1: Complete the square.

a Let $6x^2 + 11x + 3 \equiv A(x + B)^2 + C$
$$\equiv Ax^2 + 2ABx + AB^2 + C$$

Coefficient of x^2: $\qquad 6 = A$

Coefficient of x: $\qquad 11 = 2AB \Rightarrow B = \frac{11}{12}$

Constant term: $\qquad 3 = AB^2 + C \Rightarrow C = 3 - 6(\frac{11}{12})^2 = -\frac{49}{24}$

$6x^2 + 11x + 3 \equiv 6(x + \frac{11}{12})^2 - \frac{49}{24}$

Note:
$3 - 6(\frac{11}{12})(\frac{11}{12})$
$= 3 - \frac{121}{24}$
$= \frac{72}{24} - \frac{121}{24}$
$= -\frac{49}{24}$

Step 2: Make x the subject.

b Rearrange: $\qquad 6(x + \frac{11}{12})^2 = \frac{49}{24}$

Divide by 6: $\qquad (x + \frac{11}{12})^2 = \frac{49}{144}$

Take the square root of each side:

$$x + \frac{11}{12} = \pm\sqrt{\frac{49}{144}}$$
$$x + \frac{11}{12} = \pm\frac{7}{12}$$
$$x = -\frac{11}{12} \pm \frac{7}{12}$$

Either $\quad x = -\frac{11}{12} + \frac{7}{12} \quad$ or $\quad x = -\frac{11}{12} - \frac{7}{12}$

$\qquad\quad = -\frac{1}{3} \qquad\qquad\qquad = -\frac{3}{2}$

Tip:
Bring x to one side and all other terms to the other side.

Tip:
Remember to include $\pm$ when taking the square root.

Using the quadratic formula

If $ax^2 + bx + c = 0$ then

$$x = \frac{-b \pm \sqrt{b^2 - 4ac}}{2a}$$

You must memorise this formula for your examination.

Note:
The formula is derived by completing the square on $ax^2 + bx + c = 0$.

Example 1.13 Use the quadratic formula to solve $2x^2 + x - 4 = 0$, leaving your answers in surd form.

Step 1: Identify a, b and c.

$a = 2$, $b = 1$, $c = -4$

Step 2: Substitute into the quadratic formula and evaluate.

$$x = \frac{-b \pm \sqrt{b^2 - 4ac}}{2a}$$

$$= \frac{-1 \pm \sqrt{1^2 - 4(2)(-4)}}{2 \times 2}$$

$$= \frac{-1 \pm \sqrt{33}}{4}$$

$$x = \frac{-1 + \sqrt{33}}{4} \quad \text{or} \quad \frac{-1 - \sqrt{33}}{4}$$

Note:
These are exact answers. In this case, the values given by a calculator would be approximations.

Number of real roots of a quadratic equation

The term under the square root in the quadratic formula is the **discriminant**, $b^2 - 4ac$. Its value can be used to deduce the number of roots of the quadratic equation $ax^2 + bx + c = 0$.

If $b^2 - 4ac > 0$, then $\sqrt{b^2 - 4ac}$ can be calculated and the quadratic formula will give **two different values** for x.
The quadratic equation has **two real distinct roots**.

If $b^2 - 4ac = 0$, then $\sqrt{b^2 - 4ac} = 0$ and the quadratic formula will give **only one value** for x.
The quadratic equation has one real root, a **repeated root**.

If $b^2 - 4ac < 0$, then there are no real values of $\sqrt{b^2 - 4ac}$.
The quadratic equation has **no real roots**.

Recall:
The discriminant (Section 1.3).

Tip:
Learn these conditions.

Note:
See Section 1.8 (Inequalities) for more on these conditions.

Example 1.14 The equation $2x^2 - 3x + 3k = 0$ has one real root. Find the value of k.

Step 1: Identify a, b and c. $a = 2$, $b = -3$, $c = 3k$

Step 2: Find $b^2 - 4ac$ and use the appropriate condition for the discriminant.

$$b^2 - 4ac = (-3)^2 - 4(2)(3k)$$
$$= 9 - 24k$$

One real root $\Rightarrow b^2 - 4ac = 0$

$$9 - 24k = 0$$
$$k = \frac{9}{24}$$
$$= \frac{3}{8}$$

SKILLS CHECK   **1C: Quadratic equations**

1 Solve the following quadratic equations:

   a $(x - 2)(x + 3) = 0$ **b** $(1 - 4x)(3 + 2x) = 0$ **c** $4x(x + 5) = 0$

2 Solve the following equations, using the method of factorisation.

   a $x^2 + 6x + 5 = 0$ **b** $x^2 - 11x + 24 = 0$ **c** $x^2 - 6x = 0$

   d $x^2 - 5x - 6 = 0$ **e** $x^2 - x - 6 = 0$ **f** $x^2 - 36 = 0$

 3 The function f is defined for all x by $f(x) = x^2 + 3x - 5$.

   a Express $f(x)$ in the form $(x + P)^2 + Q$.

   b Hence, or otherwise, solve the equation $f(x) = 0$, giving your answers in surd form.

4 **a** Write $2x^2 - 3x - 2$ in the form $A(x + B)^2 + C$.

   b Hence solve $2x^2 - 3x - 2 = 0$.

   c Check your answers by solving $2x^2 - 3x - 2 = 0$ by the method of factorisation.

5 Use the quadratic formula to solve the equation $5x^2 + 3x - 3 = 0$, leaving your answers in surd form.

 6 **a** Solve $(2x - 3)^2 = 25$.

   b Solve $(2x - 3)^2 = 2x$, expressing your answers in surd form.

7 Find the exact solutions of the quadratic equation $3x^2 + 2x - 4 = 0$.

8 By calculating the discriminant, find the number of real roots of each of the following quadratic equations:

a $x^2 - 3x + 1 = 0$ **b** $2x^2 - 3x - 1 = 0$ **c** $4x^2 - 4x + 1 = 0$ **d** $5x + x^2 - 3 = 0$

 9 Find the discriminant of $3x^2 - 2x + 5$ and hence show that $3x^2 - 2x + 5 = 0$ has no real solutions.

SKILLS CHECK **1C EXTRA** is on the CD

1.7 Simultaneous equations

Simultaneous equations, e.g. one linear and one quadratic; analytical solution by substitution.

Linear simultaneous equations

In a pair of simultaneous equations there are two unknowns, for example x and y.

Consider these simultaneous equations.

$2y + 3x = 18$ ①
$5y - x = 11$ ②

To **solve** them you have to find a value of x and a value of y that satisfy *both* equations. The most common methods to use are elimination and substitution.

Elimination method

When the numerical coefficients of one of the unknowns are the same, *eliminate* that unknown either by adding or subtracting the equations.

Step 1: Make the coefficients of one of the unknowns the same then add or subtract to eliminate the unknown.

Step 2: Substitute the value found into one of the original equations.

② × 3 $15y - 3x = 33$ ③
                $ 2y + 3x = 18$ ①
③ + ① $\overline{ 17y = 51}$
                $ y = 3$

Substituting $y = 3$ into equation ①:
$6 + 3x = 18$
$ 3x = 12 \Rightarrow x = 4$
The solution is $x = 4$, $y = 3$.

> **Note:**
> If the terms containing the unknown you want to eliminate have the same sign, subtract. If the signs are different, add.

> **Note:**
> The lines $2y + 3x = 18$ and $5y - x = 11$ intersect at $(4, 3)$. See Sections 1.14 and 2.6 for more on graphical interpretations.

Substitution method

Use one equation to express one of the unknowns in terms of the other and then *substitute* for it in the other equation.

Step 1: Express one unknown in terms of the other.

Step 2: Substitute it into the other equation and solve.

Write x in terms of y using equation ②:
$x = 5y - 11$

Substituting for x in equation ①:
$2y + 3(5y - 11) = 18$
$2y + 15y - 33 = 18$
$17y = 51 \Rightarrow y = 3$

> **Tip:**
> Although it does not matter whether you write x in terms of y, or y in terms of x, try to avoid expressions with fractions where possible.

Step 3: Substitute the value found into one of the original equations.

To find x, proceed as in the elimination method.

11

One linear and one quadratic equation

Example 1.15   Solve the simultaneous equations.

$2x + y = 6$ ①

$y = 4 + x - x^2$ ②

Step 1: Substitute for one of the unknowns from one equation into the other.

From ① $y = 6 - 2x$

Substituting into ② $6 - 2x = 4 + x - x^2$

Step 2: Solve the resulting equation.

Rearranging $x^2 - 3x + 2 = 0$

$$(x - 1)(x - 2) = 0$$

Either $x - 1 = 0$ or $x - 2 = 0$

$\qquad\quad x = 1 \qquad\qquad x = 2$

Recall:
Solving quadratic equations (Section 1.6).

Step 3: Substitute the values found in turn into one of the original equations.

Substituting for x into ①:

When $x = 1$, $2 + y = 6$ $\Rightarrow$ $y = 4$

When $x = 2$, $4 + y = 6$ $\Rightarrow$ $y = 2$

Step 4: State the solutions in pairs.

The solutions are $x = 1, y = 4$ or $x = 2, y = 2$.

Note:
The line and the curve intersect at (1, 4) and (2, 2) (see Section 1.14).

SKILLS CHECK **1D: Simultaneous equations**

1 Solve these linear simultaneous equations by the method of elimination:

a $8x + 5y = 37$
$2x - 5y = 3$

b $3x + 2y = 13$
$x + y = 5$

c $3a - 2b = 13$
$5a - 7b = 29$

2 Solve these linear simultaneous equations by the method of substitution:

a $y = x + 10$
$3y + 2x = 0$

b $a - 2b = 5$
$2a = 5b + 7$

c $p = 3 - 4q$
$p = q - 12$

 3 **a** Show that the equation $\dfrac{x}{3} + \dfrac{y}{2} = 1$ can be written in the form $2x + 3y = 6$.

b Solve these simultaneous equations for x and y:

$\dfrac{x}{3} + \dfrac{y}{2} = 1$; $x + 3y = 2$

4 Solve for x and y:

a $y + x = 10$
$y = x^2 + 3x - 2$

b $y = 3x$
$xy + x = 2$

 c $y = x + 1$
$x^2 + y^2 = 1$

 5

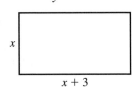

The area of this rectangle is y cm^2 and the perimeter is y cm. Find the dimensions of the rectangle.

6 A curve has equation $y = ax^2 + bx$, where a and b are integers.

a The point (3, 21) lies on the curve. Show that $7 = 3a + b$.

b The point (2, 8) also lies on the curve. Find another linear equation satisfied by a and b.

c Solve the two equations simultaneously to find the equation of the curve.

SKILLS CHECK **1D EXTRA is on the CD**

Operations on inequalities

If you perform these operations on both sides of an inequality, the inequality is unaffected:

- add a number
- subtract a number
- multiply by a *positive* number
- divide by a *positive* number.

Remember that the inequality is reversed if you do the following:

- multiply by a *negative* number
- divide by a *negative* number.

Never multiply or divide an inequality by *an unknown quantity*, as you do not know whether it is positive or negative!

> **Tip:**
> Avoid these two operations if possible, as in Example 1.16b.

Linear inequalities

Example 1.16 Find the values of x for which

a $3x - 4 > 2$ **b** $4 - x \geqslant 1$ **c** $3 < 2x + 1 < 9$

Step 1: Simplify the inequality.

Step 2: Solve the inequality.

a
$$3x - 4 > 2$$
$$(+4) \quad 3x > 6$$
$$(\div 3) \quad x > 2$$

b
$$4 - x \geqslant 1$$
$$(+x) \quad 4 \geqslant x + 1$$
$$(-1) \quad 3 \geqslant x$$
$$x \leqslant 3$$

> **Tip:**
> Adding x to both sides avoids dealing with $-x$. .

c
$$3 < 2x + 1 < 9$$
$$(-1) \quad 2 < 2x < 8$$
$$(\div 2) \quad 1 < x < 4$$

Example 1.17 Solve the inequality $6(a + 3) > 8 - 2(a + 1)$.

Step 1: Expand the brackets.
$$6a + 18 > 8 - 2a - 2$$
$$6a + 18 > 6 - 2a$$

Step 2: Simplify and solve the inequality.
$$8a + 18 > 6$$
$$8a > -12$$
$$a > -\frac{3}{2}$$

> **Tip:**
> Take care with the signs when expanding.

Quadratic inequalities

Inequalities such as $x^2 < 9$ and $x^2 \geqslant 25$ are the simplest quadratic inequalities to solve.

a To solve $x^2 < 9$, first consider $x^2 = 9$.

$x^2 = 9$ when $x = 3$ or $x = -3$.

For any value between -3 and 3, $x^2 < 9$.

So the range of values of x for which $x^2 < 9$ is $-3 < x < 3$.

Tip:
In a) the values are 'sandwiched' between -3 and 3, so the solution should be written in one inequality.

b To solve $x^2 \geqslant 25$, first consider $x^2 = 25$.

$x^2 = 25$ when $x = 5$ or $x = -5$.

If $x \leqslant -5$, $x^2 \geqslant 25$. Also if $x \geqslant 5$, $x^2 \geqslant 25$.

So the values of x for which $x^2 \geqslant 25$ are $x \leqslant -5$ or $x \geqslant 5$.

Tip:
In b) the values are outside the 'sandwich', so give two separate inequalities.

This method can be applied to more complicated quadratic inequalities by completing the square.

Recall:
Completing the square (Section 1.5).

Example 1.18

a Express $x^2 - 6x + 7$ in the form $(x + a)^2 + b$.

b Find the range of values of x for which $x^2 - 6x + 7 < 0$.

Step 1: Complete the square.

a $x^2 - 6x + 7 = (x - 3)^2 - 9 + 7$
$\qquad\qquad\qquad = (x - 3)^2 - 2$

b $\qquad x^2 - 6x + 7 < 0$

Step 2: Solve the inequality using the completed square format.

$\Rightarrow (x - 3)^2 - 2 < 0$
$\qquad (x - 3)^2 < 2$
$\qquad -\sqrt{2} < x - 3 < \sqrt{2}$
$(+3) \quad 3 - \sqrt{2} < x < 3 + \sqrt{2}$

So $x^2 - 6x + 7 < 0$ when $3 - \sqrt{2} < x < 3 + \sqrt{2}$.

Tip:
Substitute y for $(x - 3)$ and use the fact that if $y^2 < 2$, then $-\sqrt{2} < y < \sqrt{2}$.

Recall:
Surds (Section 1.1).

Another method is to use a sketch. This is illustrated in the Example 1.19.

Example 1.19 It is given that $f(x) = x^2 + 2x - 8$.

a Solve $f(x) = 0$.

b Sketch $y = f(x)$.

c Hence solve $x^2 + 2x - 8 \geqslant 0$.

Note:
Sketching a quadratic curve is described more fully in Section 1.13.

Step 1: Solve the quadratic equation.

a $f(x) = 0$ when $x^2 + 2x - 8 = 0$
$\qquad\qquad\qquad (x - 2)(x + 4) = 0$
$\Rightarrow \quad x = 2, -4$.

The curve $y = f(x)$ crosses the x-axis at $(2, 0)$ and $(-4, 0)$.

Recall:
Solution of quadratic equations (Section 1.6).

Step 2: State where the curve crosses the x-axis.

Step 3: Sketch the curve.

b

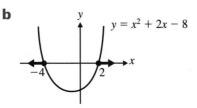

Recall:
The coefficient of x^2 is positive, so the parabola is $\cup$-shaped (Section 1.2).

Tip:
You can use 'filled in' circles to show that x can be 2 or -4.

Step 4: Indicate the possible x-values on the sketch and solve $f(x) \geqslant 0$.

c $f(x) \geqslant 0$ when the curve is on or above the x-axis, so from the sketch, $f(x) \geqslant 0$ when $x \leqslant -4$, $x \geqslant 2$.

Tip:
Write this solution in two separate inequalities.

Application of inequalities to roots of equations

You may need to solve a quadratic inequality in questions about roots of equations, as in the following example.

Example 1.20 Find the values of k for which $2x^2 - kx + 2 = 0$ has no real roots.

Step 1: Identify a, b and c.

Comparing $2x^2 - kx + 2$ with $ax^2 + bx + c$,

$a = 2, b = -k, c = 2$

If the equation has no real roots, then
$$b^2 - 4ac < 0$$

Step 2: Find $b^2 - 4ac$, and use the appropriate condition for the discriminant.

$\Rightarrow \quad (-k)^2 - 4 \times 2 \times 2 < 0$
$$k^2 - 16 < 0$$

Step 3: Solve the inequality in k by sketching $y = f(k)$ and finding the values of k when the curve is below the k-axis.

Let $f(k) = k^2 - 16$
$$= (k - 4)(k + 4)$$
$f(k) = 0$ when $k = 4$ or -4

The graph of $y = f(k)$ goes through $(-4, 0)$ and $(4, 0)$.

From the sketch, $f(k) < 0$ when $-4 < k < 4$.

So $2x^2 - kx + 2 = 0$ has no real roots when $-4 < k < 4$.

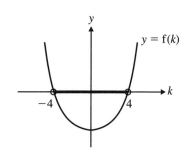

> **Recall:**
> Conditions to be satisfied by the discriminant (Section 1.6).

> **Tip:**
> You can use 'open' circles to show that k cannot take the values 4 and -4.

> **Note:**
> This solution can be written in one statement.

SKILLS CHECK   1E: Inequalities

1 Solve these linear inequalities.

   a $4x - 5 > 7$ **b** $3 - 5x < 8$ **c** $2(3x - 1) \geqslant 3(x + 8)$

   d $5y - (4 + y) > 0$ **e** $4 < \dfrac{2x}{3}$ **f** $\dfrac{1}{2}x + \dfrac{3}{4} \leqslant 8$

2 Solve these inequalities:

   a $5 < 2x - 1 < 17$ **b** $-3 \leqslant \dfrac{x}{2} \leqslant 5$

 3 The solution of the inequality $\sqrt{3}(x + \sqrt{3}) > 6$ is $x > \sqrt{a}$. Find the value of a.

4 Solve these quadratic inequalities.

   a $y^2 > 4$ **b** $x^2 \leqslant 49$ **c** $x^2 \geqslant 5$

   d $2x^2 < 18$ **e** $(x - 1)^2 > 4$ **f** $(x + 2)^2 \leqslant 5$

5 **a** Express $x^2 + 4x - 5$ in the form $(x + p)^2 + q$, finding the values of the constants p and q.

   b Find the values of x for which $x^2 + 4x \geqslant 5$.

6 Solve these quadratic inequalities.

   a $(x + 4)(x - 3) < 0$ **b** $3(2x + 5)(3x - 2) \geqslant 0$ **c** $(4 - x)(5 + x) \leqslant 0$

   d $p^2 + 7p + 10 < 0$ **e** $2x^2 + x \geqslant 6$

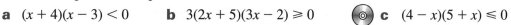

15

7 a The quadratic equation $x^2 - 4x - 6 = 0$ has solutions $x = p \pm \sqrt{q}$ where p and q are integers. Find the values of p and q.

b Sketch $y = x^2 - 4x - 6$.

c Solve the inequality $x^2 - 4x - 6 < 0$.

8 a Find the values of k for which the equation $3x^2 - kx + 3 = 0$ has no real roots.

b The equation $kx^2 - 8x + k = 0$ has real roots. Show that $-4 \leqslant k \leqslant 4$.

9 a Find the discriminant of the quadratic expression $2x^2 - kx + 2$.

b Find the values of k for which $2x^2 - kx + 2 = 0$ has two distinct real roots.

SKILLS CHECK **1E EXTRA** is on the CD

1.9 Algebraic manipulation of polynomials

Algebraic manipulation of polynomials, including expanding brackets and collecting like terms.

A **polynomial in x** is an expression with positive integer powers of x, for example $4x^3 + 2x^2 - 7x + 6$. The **degree** of a polynomial is the highest power of x, so this polynomial has degree 3.

> **Note:**
> The usual convention is to write the polynomial with the highest power first.

Example 1.21 Expand and simplify the polynomials.

Step 1: Expand brackets.

Step 2: Collect like terms.

a $2x(x^2 + 2x - 1) + (x - 3)(x + 2)$
$= 2x^3 + 4x^2 - 2x + x^2 + 2x - 3x - 6$
$= 2x^3 + 5x^2 - 3x - 6$

b $(x - 2)(x + 1)(x - 3)$
$= (x^2 - x - 2)(x - 3)$
$= x^3 - 3x^2 - x^2 + 3x - 2x + 6$
$= x^3 - 4x^2 + x + 6$

> **Tip:**
> Expand two of the brackets, then multiply by the third.

c $(x^2 + x - 3)(2x^2 - 4x + 2)$
$= 2x^4 - 4x^3 + 2x^2 + 2x^3 - 4x^2 + 2x - 6x^2 + 12x - 6$
$= 2x^4 - 2x^3 - 8x^2 + 14x - 6$

> **Tip:**
> Multiplying two brackets, each with three terms, will give nine terms, some of which will combine. Don't try to do this mentally.

Example 1.22 It is given that
$(ax + b)(2x^2 - 3x + 5) \equiv 4x^3 + cx^2 + 7x - 5.$

Find the values of a, b and c.

> **Note:**
> This is an identity, true for all values of x.

Step 1: Expand the left-hand side to get the x^3 and constant terms.

Step 2: Equate terms to find a and b.

Consider x^3 and constant terms:
$2ax^3 + \cdots + 5b \equiv 4x^3 + cx^2 + 7x - 5$

| Equate x^3 terms: | $2a = 4$ | $\Rightarrow a = 2$ |
| Equate constants: | $5b = -5$ | $\Rightarrow b = -1$ |

> **Note:**
> There is only one way to get the x^3 term and this will give the value of a. Similarly, equating the constant terms will give b.

Step 3: Expand to get the x^2 term and solve for c.

Consider x^2 terms:
$\cdots + 2bx^2 - 3ax^2 + \cdots \equiv 4x^3 + cx^2 + 7x - 5$

Equate x^2 terms: $\qquad 2b - 3a = c$
$$-2 - 6 = c \Rightarrow c = -8$$

So $a = 2$, $b = -1$ and $c = -8$.

1 Expand and simplify the polynomials:

a $2x(x-2) - (3x-1)(x+5)$ **b** $3x^2(x+4) + (x^2 - 2x + 1)(x+4)$

c $(x+3)(x-5)(x-3)$ **d** $(2x+1)(x-3)(x+4)$

2 Find the values of the letters:

a $(ax+3)(x-4) \equiv 2x^2 - bx - 12$ **b** $(ax^2 + bx + 3)(2x - 1) \equiv 6x^3 - cx^2 + 7x - 3$

 c $(ax+2)(bx-1) \equiv 6x^2 + x - c$

1.10 The Factor Theorem

Use of the Factor Theorem.

For a polynomial P(x), the **Factor Theorem** states that if a is a root of the equation P(x) = 0, then $(x - a)$ is a factor of P(x), and vice versa.

This can be written

$$P(a) = 0 \Leftrightarrow (x - a) \text{ is a factor of P}(x).$$

For example,

- if P(3) = 0, then $(x - 3)$ is a factor of P(x)

- if P(−4) = 0, then $(x + 4)$ is a factor of P(x)

- if $(x - 5)$ is a factor of P(x), then P(5) = 0

- if $(x + 6)$ is a factor of P(x), then P(−6) = 0.

> **Note:**
> The symbol ⇔ means that the statement is true when read from left to right, or from right to left.

The Factor Theorem can be used to identify factors of P(x) and hence solve P(x) = 0, as in the following example.

Example 1.23 It is given that $P(x) = x^3 + 4x^2 + x - 6$.

a Find the value of P(1).

b Use the Factor Theorem to write down a factor of P(x).

c Express P(x) as a product of three linear factors.

d Hence solve the equation $x^3 + 4x^2 + x - 6 = 0$.

Step 1: Substitute the value into P(x).

a $P(1) = 1^3 + 4 \times 1^2 + 1 - 6 = 1 + 4 + 1 - 6 = 0$

Step 2: Use the Factor Theorem.

b $P(1) = 0 \Rightarrow (x - 1)$ is a factor of P(x).

c $P(x) = (x - 1)(ax^2 + bx + c)$

so $x^3 + 4x^2 + x - 6 \equiv (x - 1)(ax^2 + bx + c)$

Step 3: Write down the linear factor and find the quadratic factor.

Equate x^3 terms:	$x^3 = ax^3$	$\Rightarrow a = 1$
Equate constants:	$-6 = -c$	$\Rightarrow c = 6$
Equate x^2 terms:	$4x^2 = bx^2 - ax^2$	$\Rightarrow 4 = b - a \Rightarrow b = 5$

Hence $P(x) = (x - 1)(x^2 + 5x + 6)$.

> **Note:**
> You could use algebraic division (see Section 1.12).

d $P(x) = 0 \Rightarrow (x - 1)(x^2 + 5x + 6) = 0$

$(x - 1)(x + 2)(x + 3) = 0$

So $x = 1$, $x = -2$, $x = -3$.

If, in the above example, you had not been given a hint to find the first factor, then a good strategy would be to try numbers that are factors of the constant term. In this case, try factors of -6, for example $1, -1, 2, -2, 3, -3$.

Factors of a cubic polynomial

A cubic polynomial, P(x), can be factorised into a linear and a quadratic factor. You may then be able to factorise further.

So P(x) will have one of the following:

- three distinct linear factors,

- repeated linear factors,

- a linear factor and a quadratic factor that cannot be factorised in real numbers.

Solving a cubic equation

When trying to solve P(x) = 0, factorise P(x) into a linear and a quadratic factor. Then try to factorise the quadratic factor.

If it does not appear to factorise, check the discriminant $b^2 - 4ac$, remembering that

Recall:
The discriminant (Sections 1.3 and 1.6).

- if $b^2 - 4ac \geqslant 0$, the quadratic equation can be solved by completing the square or using the quadratic formula,

- if $b^2 - 4ac < 0$, the quadratic equation will have no real solutions, and P(x) = 0 will have only one root.

1.11 The Remainder Theorem

Use of the Remainder Theorem.

The **Remainder Theorem** states that if a polynomial P(x) is divided by a linear term $(x - a)$, the remainder will be P(a).

The Remainder Theorem is sometimes written

$$\frac{P(x)}{x - a} = Q(x) + \frac{P(a)}{x - a}$$

where Q(x) is another polynomial of one degree less than P(x).

Tip:
For Core 1, P(x) will be restricted to quadratic and cubic polynomials.

Note:
The Factor Theorem is a special case of the Remainder Theorem, because if $(x - a)$ is a factor of P(x), then P(a) = 0, indicating that $x = a$ is a root of P(x) = 0.

Example 1.24 Find the remainder when $P(x) = x^3 - 4x^2 - 5x + 7$ is divided by $(x + 2)$.

Step 1: Substitute $x = -2$ into the polynomial.

$P(-2) = (-2)^3 - 4(-2)^2 - 5(-2) + 7 = -8 - 16 + 10 + 7 = -7$

The remainder is -7.

Tip:
Be careful with signs.

1.12 Algebraic division

Simple algebraic division.

Note:
Algebraic division has the same structure as traditional long division with numbers.

 Example 1.25 Divide $x^3 + 2x^2 - 3x + 7$ by $x - 2$.

There are 12 steps to this process. These are outlined on a PowerPoint slide Example 1.25 which can be found on the CD.

$$\begin{array}{r} x^2 + 4x + 5 \\ x - 2{\overline{)x^3 + 2x^2 - 3x + 7}} \\ \underline{x^3 - 2x^2} \\ 4x^2 - 3x \\ \underline{4x^2 - 8x} \\ 5x + 7 \\ \underline{5x - 10} \\ +17 \end{array}$$

$$(x^3 + 2x^2 - 3x + 7) \div (x - 2) = x^2 + 4x + 5 + \frac{17}{x - 2}$$

Note:
The remainder 17 can be confirmed by substituting $x = 2$ into the polynomial.
$P(2) = 8 + 8 - 6 + 7 = 17$

 Example 1.26 $P(x) = x^3 + 3x^2 - 4x - 12$

Factorise P(x) fully, given that $x = -3$ is a root of P(x).

$x = -3$ is a root of $P(x) = 0 \Rightarrow (x + 3)$ is a factor of P(x).

Step 1: Divide the expression by the known linear factor.

$$\begin{array}{r} x^2 \qquad - 4 \\ x + 3{\overline{)x^3 + 3x^2 - 4x - 12}} \\ \underline{x^3 + 3x^2} \\ 0 - 4x - 12 \\ \underline{- 4x - 12} \\ 0 \end{array}$$

Step 2: Factorise the dividend.
A full list of the 11 steps are outlined on a PowerPoint slide Example 1.26 which can be found on the CD.

$x^2 - 4 = (x + 2)(x - 2)$
$x^3 + 3x^2 - 4x - 12 = (x + 3)(x - 2)(x + 2)$

Note:
Using algebraic division is an alternative to the method of identities described in Example 1.23.

Recall:
$x^2 - 4$ factorises using the difference of two squares.

Example 1.27 Use the Factor Theorem to find a linear factor of P(x) where $P(x) = 2x^3 - 5x^2 - x + 6$. Hence express P(x) as a product of three linear factors.

Step 1: Find a value of x such that $P(x) = 0$.

Try $x = 1$: $P(1) = 2 - 5 - 1 + 6 = 2 \neq 0$. Hence $(x - 1)$ is not a factor.

Try $x = -1$: $P(-1) = -2 - 5 + 1 + 6 = 0$.

Step 2: Find the quadratic factor.

Hence $(x + 1)$ is a factor of P(x).

Divide P(x) by $(x + 1)$:

$$\begin{array}{r} 2x^2 - 7x + 6 \\ x + 1{\overline{)2x^3 - 5x^2 - \ x + 6}} \\ \underline{2x^3 + 2x^2} \\ -7x^2 - x \\ \underline{-7x^2 - 7x} \\ 6x + 6 \\ \underline{6x + 6} \\ 0 \end{array}$$

Tip:
Try factors of 6: 1, −1, …

Step 3: Find the three linear factors. $2x^2 - 7x + 6 = (2x - 3)(x - 2)$

Hence $P(x) = (x + 1)(2x - 3)(x - 2)$.

Example 1.28   When $f(x) = ax^3 + bx^2 + 4x - 6$ is divided by $(x - 1)$ the remainder is -1. When $f(x)$ is divided by $x - 3$ the remainder is 42.

Find the values of a and b.

Step 1: Use the Remainder Theorem twice.   By the Remainder Theorem $f(1) = -1$.

Hence $a(1)^3 + b(1)^2 + 4(1) - 6 = -1$

$$a + b = 1 \qquad ①$$

Also by the Remainder Theorem $f(3) = 42$.

Hence $a(3)^3 + b(3)^2 + 4(3) - 6 = 42$

$$27a + 9b = 36 \qquad ②$$

Step 2: Solve the simultaneous equations formed.   Solve the simultaneous equations ① and ②:

$① × 27$ $\qquad\qquad 27a + 27b = 27 \qquad ③$

$③ - ②$ $\qquad\qquad\qquad 18b = -9$

$$b = -\tfrac{1}{2}$$

Substitute into ① $\qquad\qquad\qquad a = \tfrac{3}{2}$

SKILLS CHECK 1G: Factor and Remainder Theorems

1 For the following polynomials, write down the roots of $P(x) = 0$.

a $P(x) = (2x - 1)(x - 2)(3x + 1)$ **b** $P(x) = (x^2 - 4)(x^2 - 9)$

c $P(x) = (4x - 1)(3x + 4)(x - 2)$ **d** $P(x) = (x + 3)(x - 1)^2$

2 Find the roots of the following, giving your answer in surd form, if appropriate.

a $x^3 - 7x + 6 = 0$ **b** $x^3 + x^2 - 10x - 6 = 0$ **c** $x^3 + 6x^2 + 11x + 6 = 0$

3 Find the remainder when the given polynomial is divided by the given linear term.

a $x^3 - 3x^2 + 2x - 1$ divided by $(x - 2)$.

b $x^3 + 7x^2 + 8x + 10$ divided by $(x + 1)$.

c $x^3 + 3x^2 - 4$ divided by $(x - 1)$.

Explain the significance of the result in part **c**.

4 Divide the given polynomial by the given linear term.

 a $x^3 - 6x^2 + 4x - 3$ divided by $(x - 3)$.

b $x^3 + 6x^2 - x + 1$ divided by $(x + 1)$.

c $2x^3 + x^2 - 4x + 5$ divided by $(x - 2)$.

5 Find the roots of the equation $x^3 - 6x^2 + x + 14 = 0$ given that 2 is a root. Give your answers in surd form if appropriate.

6 Use the Factor Theorem to find a linear factor of $P(x)$ where $P(x) = x^3 + 3x^2 + 3x + 1$. Hence express $P(x)$ as a product of three linear factors.

7 When the polynomial $f(x) = ax^2 + bx - 1$ is divided by $(x - 2)$ the remainder is 15.
When $f(x)$ is divided by $(x + 1)$ the remainder is –3.
Find the values of a and b.

 8 When the polynomial $f(x) = ax^3 + bx^2 + 3x + 4$ is divided by $(x + 1)$ the remainder is -4.
When $f(x)$ is divided by $(x - 2)$ the remainder is 38.
Find the values of a and b.

SKILLS CHECK **1G EXTRA** is on the CD

1.13 Sketching curves

Graphs of functions; sketching curves defined by simple equations.

To sketch a curve, identify

- what general shape the curve takes

- where the curve crosses the x-axis, by setting $y = 0$

- where the curve crosses the y-axis, by setting $x = 0$.

> **Note:**
> You must indicate the intercepts with the axes on the sketch.

Linear functions in *x*

The function $f(x) = ax + b$, where $a \neq 0$, is a **linear** polynomial in x.
The graph of $y = ax + b$ is a straight line, with y-intercept b and gradient a.

> **Note:**
> The highest power of x is 1.

$$y = ax + b$$

> **Note:**
> For more on lines, see Chapter 2: Coordinate geometry.

> **Note:**
> See Section 2.1 for other formats of the equation of a straight line.

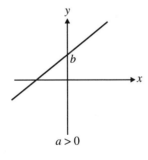

$a > 0$

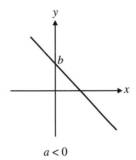

$a < 0$

Quadratic functions in *x*

The function $f(x) = ax^2 + bx + c$, where $a \neq 0$, is a **quadratic polynomial** in x and the graph of $y = f(x)$ is a **parabola**.

> **Note:**
> The highest power of x is 2.

$a > 0$

$a < 0$

> **Recall:**
> Quadratic functions (Section 1.2).

Intercepts with the axes:

$x = 0 \Rightarrow y = c$, so the curve passes through $(0, c)$.

$y = 0 \Rightarrow ax^2 + bx + c = 0$, so the intercepts on the x-axis are found by solving the quadratic equation.

> **Recall:**
> Quadratic equations (Section 1.6).

Note that the axis of symmetry of the parabola is a vertical line passing through the mid-point between the intercepts on the x-axis.

Example 1.29 It is given that $f(x) = 2x^2 + 5x - 3$.

    a Sketch $y = f(x)$, labelling the intercepts with the axes.

    b Draw the axis of symmetry on the sketch and state its equation.

Recall:
Quadratic functions (Section 1.6).

Step 1: Decide the general shape.

a $f(x)$ is a polynomial of degree 2, so the curve is a parabola.
$a > 0 \Rightarrow$ parabola is $\cup$-shaped.

Note:
The axis of symmetry and the vertex can also be found by completing the square (Section 1.5) or by calculus (Section 3.6).

Step 2: Set $y = 0$ and $x = 0$ to find the axes intercepts.

When $x = 0$, $y = -3$, so the curve goes through $(0, -3)$.

When $y = 0$, $\qquad 2x^2 + 5x - 3 = 0$

$\qquad\qquad\qquad (2x - 1)(x + 3) = 0$

$\qquad\qquad \Rightarrow \quad x = 0.5$ or $x = -3$

Step 3: Sketch the curve, marking the intercepts.

$(0.5, 0)$ and $(-3, 0)$ lie on the curve.

Mid-point of x-intercepts $= \dfrac{-3 + 0.5}{2} = -1.25$

Step 4: Draw the axis of symmetry.

b The axis of symmetry is the line $x = -1.25$.

Step 5: Find the axis of symmetry using the x-axis intercepts.

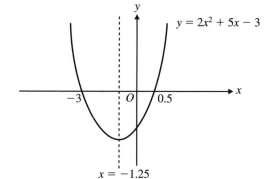

Significance of the discriminant

The discriminant can be used to determine whether or not the curve crosses or touches the x-axis.

If $b^2 - 4ac > 0$, the equation $ax^2 + bx + c = 0$ has two distinct (different) roots. These give the two x-coordinates of the points where the graph of $y = ax^2 + bx + c$ crosses the x-axis.

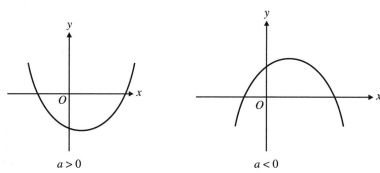

If $b^2 - 4ac = 0$, then $ax^2 + bx + c = 0$ has exactly one real root. This gives the x-coordinate where the graph of $y = ax^2 + bx + c$ touches the x-axis.

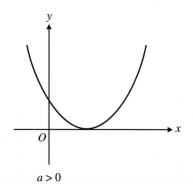

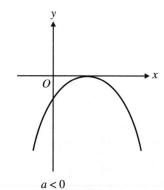

Note:
The x-axis is a tangent to the curve (see Sections 2.6 and 3.4).

If $b^2 - 4ac < 0$, then $ax^2 + bx + c = 0$ has no real roots.

This tells you that the graph of $y = ax^2 + bx + c$ does not cross or touch the x-axis.

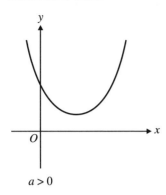
$a > 0$

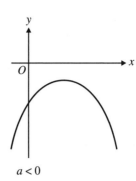
$a < 0$

Example 1.30   Show that the curve $y = 3x^2 - 2x + 5$ lies entirely above the x-axis.

When $x = 0$, $y = 5$, so the curve passes through $(0, 5)$.

When $y = 0$, $3x^2 - 2x + 5 = 0$

Step 1: Find the discriminant.
$a = 3$, $b = -2$, $c = 5$
$b^2 - 4ac = (-2)^2 - 4 \times 3 \times 5 = 4 - 60 = -56$

Step 2: Use condition on discriminant.
Since $b^2 - 4ac < 0$, the equation $3x^2 - 2x + 5 = 0$ has no real roots and the curve does not cross or touch the x-axis.

Since the curve passes through $(0, 5)$, it must lie entirely above the x-axis.

> **Note:**
> You could find the vertex by completing the square (Section 1.5) or using calculus (Section 3.6).

Cubic functions

The function $f(x) = ax^3 + bx^2 + cx + d$, where $a \neq 0$, is a **cubic** polynomial in x.

> **Note:**
> The highest power of x is 3.

Intercepts with the axes:
$x = 0 \Rightarrow y = d$, so the curve passes through $(0, d)$.
$y = 0 \Rightarrow ax^3 + bx^2 + cx + d = 0$, so to find the intercepts on the x-axis, solve the cubic equation.

> **Recall:**
> Factorising a cubic expression (Section 1.10).

In general, the graph of $y = ax^3 + bx^2 + cx + d$ can have the following shapes:

$a > 0$

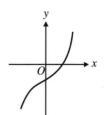

> **Note:**
> The cubic curve crosses the x-axis at least once and no more than 3 times.

$a < 0$

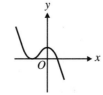

> **Note:**
> A cubic curve may have two, one or no stationary points. See Section 3.6 for more on stationary points.

Example 1.31 Sketch the curve $y = (x - 1)(2x + 1)(2 - x)$.

Step 1: Decide the general shape.

The highest power of x is 3, so the curve is a cubic.

Coefficient of x^3 is -2, so the general shape is one of those described above for $a < 0$.

Tip:
Multiplying the terms indicated gives the coefficient of x^3:
$(x - 1)(2x + 1)(2 - x)$.

Step 2: Set $y = 0$ and $x = 0$ to find the axes intercepts.

When $y = 0$, $\quad (x - 1)(2x + 1)(2 - x) = 0$

$\Rightarrow \quad x = 1, x = -\frac{1}{2}$ or $x = 2$

When $x = 0$, $y = (-1) \times 1 \times 2 = -2$

The curve goes through $(0, -2)$, $(-\frac{1}{2}, 0)$, $(1, 0)$ and $(2, 0)$.

Note:
If the cubic expression is not in factorised form, try to factorise it using the factor theorem (Section 1.10).

Step 3: Sketch the curve, marking the intercepts.

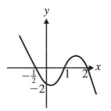

The circle

A circle with centre (a, b) and radius r has equation $(x - a)^2 + (y - b)^2 = r^2$.

If the centre is at the origin $(0, 0)$ the equation is $x^2 + y^2 = r^2$.

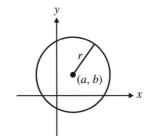

Note:
See Section 2.3 for more on the circle.

1.14 Geometrical interpretation of algebraic solution of equations

Geometrical interpretation of algebraic solution of equations and use of intersection points of graphs of functions to solve equations.

At a point where two curves, or a line and a curve, intersect or meet, the equation of each must be satisfied.

An algebraic way of finding the point of intersection is to solve the equations simultaneously.

Recall:
Simultaneous equations (Section 1.7).

The converse is also true:
If you solve two equations simultaneously you are effectively finding the point of intersection of the curves represented by the equations.

Example 1.32 The diagram shows a sketch of $y = x^2 - 9$ and $y = 4x - 12$.

The line and the curve intersect at A and B.

Find the coordinates of A and B.

Step 1: Solve the equations simultaneously.

When the line and curve intersect:

$$x^2 - 9 = 4x - 12$$
$$x^2 - 4x + 3 = 0$$
$$(x - 1)(x - 3) = 0$$
$$\Rightarrow \quad x = 1 \text{ or } x = 3$$

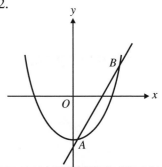

Substituting into $y = 4x - 12$:

When $x = 1$, $y = 4 \times 1 - 12 = -8$

When $x = 3$, $y = 4 \times 3 - 12 = 0$

Step 2: Use the solutions to write down the point(s) of intersection.

A is the point $(1, -8)$ and B is the point $(3, 0)$.

Note:
See Section 2.6 for more on intersections of lines and curves.

SKILLS CHECK **1H: Graphs of functions**

1 Sketch the following, showing the intercepts with the axes.

   a $y = 3x + 2$ **b** $y = 1 - 2x$ **c** $y = (x - 3)(2x + 1)$

   d $y = (3 - x)(2 + x)$ **e** $y = x^2 + 4x - 5$ **f** $y = 7 - 5x + 2x^2$

2 a Factorise $-2x^2 - 7x + 4$.

   b Sketch the graph of $y = -2x^2 - 7x + 4$, showing the intercepts with the axes.

 3 Sketch the following graphs, showing the vertex and the coordinates of any intercepts with the axes.

   a $y = (x - 3)^2$ **b** $y = (3 - x)^2$

4 It is given that $f(x) = x^2 + 2x - 24$.

   a Factorise $f(x)$ and hence solve $f(x) = 0$.

   b Write $f(x)$ in the form $(x + B)^2 + C$.

   c Using your answers from parts **a** and **b**, sketch $y = f(x)$, stating the coordinates of the intercepts with the axes and the turning point.

   d Write down the equation of the axis of symmetry.

5 Sketch the following cubic graphs, showing the intercepts with the axes.

   a $y = (x + 3)(x + 4)(x - 2)$ **b** $y = (x + 1)(6 - x)(2 - 3x)$ **c** $y = x(x - 2)^2$

 6 It is given that $f(x) = 2x^3 - x^2 - 5x - 2$.

   a Show that $(x - 2)$ is a factor of $f(x)$.

   b Factorise $f(x)$ completely.

   The graph of $y = f(x)$ crosses the y-axis at A.

   c Find the coordinates of A.

   d Find the coordinates of the points where the curve crosses the x-axis.

   e Sketch the curve $y = f(x)$.

7 Sketch the curve $x^2 + y^2 = 25$.

SKILLS CHECK **1H EXTRA is on the CD**

Knowledge of the effect of translations on graphs and their equations.

The transformation $y = x^2 + b$ has the effect of translating the graph of $y = x^2$ by $\begin{pmatrix} 0 \\ b \end{pmatrix}$.

Note:
If $b > 0$, the curve moves up. If $b < 0$, the curve moves down.

$y = x^2 + 2$ is a translation of $y = x^2$ by 2 units up.

$y = x^2 - 1$ is a translation of $y = x^2$ by 1 unit down.

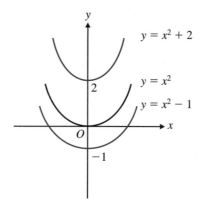

The transformation $y = (x - a)^2$ has the effect of translating the graph of $y = x^2$ by $\begin{pmatrix} a \\ 0 \end{pmatrix}$.

Note:
If $a > 0$, the curve moves right. If $a < 0$, the curve moves left.

$y = (x - 2)^2$ is a translation of $y = x^2$ by 2 units to the right.

$y = (x + 1)^2$ is a translation of $y = x^2$ by 1 unit to the left.

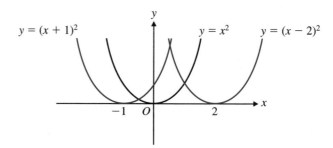

The translations can be combined, where $f(x) = (x - a)^2 + b$ has the effect of translating x^2 by $\begin{pmatrix} a \\ b \end{pmatrix}$.

Example 1.33 The translation $\begin{pmatrix} 1 \\ 3 \end{pmatrix}$ is applied to the graph $y = x^2$.

Step 1: Apply the translations
$y = f(x + b)$
$y = f(x) + a$.

State the equation of the resulting curve.

The resulting curve has equation $y = (x - 1)^2 + 3$.

Example 1.34 $f(x) = x^2 - 4x - 1$

a Write $f(x)$ in the form $y = (x - a)^2 + b$.

b Hence, or otherwise, sketch the graph $y = f(x)$, stating the coordinates of the minimum point.

Step 1: Complete the square.

a $x^2 - 4x - 1 = (x - 2)^2 - 4 - 1 = (x - 2)^2 - 5$
So $f(x) = (x - 2)^2 - 5$.

Step 2: Apply translation to sketch the curve.

b Translate $y = x^2$ by $\begin{pmatrix} 2 \\ -5 \end{pmatrix}$.

Step 3: Find the new position of the vertex.

The vertex moves from $(0, 0)$ to $(2, -5)$, so the minimum point P is $(2, -5)$.

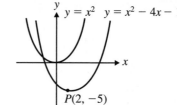

Recall:
The minimum point of $y = A(x + B)^2 + C$ is $(-B, C)$. See (Section 1.5).

Translations can also be applied to circles. In general,
$(x - a)^2 + (y - b)^2 = r^2$ is a translation of $x^2 + y^2 = r^2$ by $\begin{pmatrix} a \\ b \end{pmatrix}$.

Example 1.35 The sketch shows the circle $x^2 + y^2 = 4$.

    a Sketch the circle $(x - 2)^2 + (y + 1)^2 = 4$

    b State the coordinates of its centre.

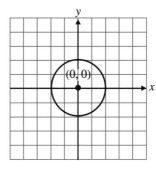

Step 1: Identify the translation. **a** Translate by $\begin{pmatrix} 2 \\ -1 \end{pmatrix}$.

Step 2: Sketch the graph in the new position. **b** The centre of the circle is at $(2, -1)$.

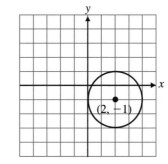

> **Note:**
> The effect of $(x - 2)^2$ is to translate the circle by $\begin{pmatrix} 2 \\ 0 \end{pmatrix}$. The effect of $(y + 1)^2$ is to translate the circle by $\begin{pmatrix} 0 \\ -1 \end{pmatrix}$.

SKILLS CHECK **1I: Translations**

1 For each of the following curves
  i describe the geometrical transformation by which the curve can be obtained from the parabola with equation $y = x^2$,
  ii sketch the curve, stating the coordinates of the vertex and the y-intercept.

  a $y = x^2 + 3$ **b** $y = x^2 - 2$ **c** $y = x^2 + 1$

  d $y = (x + 2)^2$ **e** $y = (x - 1)^2$ **f** $y = (x + 4)^2$

  g $y = (x + 3)^2 + 1$ **h** $y = (x - 2)^2 - 1$ **i** $y = (x + 1)^2 - 2$

2 For each of the following circles
  i describe the geometrical transformation by which the circle can be obtained from the circle with equation $x^2 + y^2 = r^2$,
  ii state the coordinates of the centre of the circle and the radius.

  a $(x + 2)^2 + y^2 = 4$ **b** $x^2 + (y - 2)^2 = 25$

  c $(x + 1)^2 + (y - 3)^2 = 9$ **d** $(x - 2)^2 + (y + 1)^2 = 4$

3 State the equations of each of the following circles.

  a

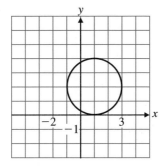

  b

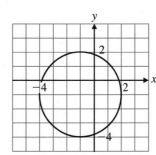

4 a Given that $f(x) = x^2 - 6x + 10$, express $f(x)$ in the form $y = (x + a)^2 + b$.

b The curve $y = x^2 - 6x + 10$ is a translation of $y = x^2$. Describe the translation and state the minimum point of the curve $y = x^2 - 6x + 10$.

c By considering the graph of $y = x^2 - 6x + 10$, explain why the equation $x^2 - 6x + 10 = 0$ has no roots.

SKILLS CHECK **1I EXTRA is on the CD**

Examination practice   Algebra

1 Express each of the following in the form $p + q\sqrt{3}$:

a $(2 + \sqrt{3})(5 - 2\sqrt{3})$;

b $\dfrac{26}{4 - \sqrt{3}}$. [AQA (B) June 2002]

 2 a Express $\dfrac{\sqrt{2} + 1}{\sqrt{2} - 1}$ in the form $a\sqrt{2} + b$, where a and b are integers.

b Solve the inequality

$$\sqrt{2}(x - \sqrt{2}) < x + 2\sqrt{2}.$$ [AQA (A) Jan 2003]

3 Solve the following inequalities:

a $5(y + 5) < 7 - 3(y + 2)$;

b $x^2 + 4x - 12 > 0$. [AQA (B) Nov 2002]

4 a Express each of the following in the form $k\sqrt{5}$:

i $\sqrt{45}$ **ii** $\dfrac{20}{\sqrt{5}}$.

b Hence write $\sqrt{45} + \dfrac{20}{\sqrt{5}}$ in the form $n\sqrt{5}$, where n is an integer. [AQA (B) June 2003]

5 Solve the simultaneous equations:

$$y = 6 - 2x$$
$$xy + x = 3.$$ [AQA (B) Jan 2002]

6 Solve the simultaneous equations:

$$y = 2 - x$$
$$x^2 + 2xy = 3.$$ [AQA (A) Jan 2002]

7 Find the values of x and y that satisfy the simultaneous equations:

$$y = 2 - x^2$$
$$x + 2y = 1.$$ [AQA (A) May 2002]

8 a Express $x^2 + 4x - 5$ in the form $(x + a)^2 + b$, finding the values of the constants a and b.

b Find the values of x for which $x^2 + 4x - 5 > 0$. [AQA (A) June 2001]

9 a i Solve $2x^2 + 8x + 7 = 0$, giving your answers in surd form.

   ii Hence solve $2x^2 + 8x + 7 > 0$.

b Express $2x^2 + 8x + 7$ in the form $A(x + B)^2 + C$, where A, B and C are constants.

c i State the minimum value of $2x^2 + 8x + 7$.

   ii State the value of x which gives this minimum value. [AQA (A) May 2002]

10 a i Express $x^2 + 12x + 11$ in the form $(x + a)^2 + b$, finding the values of a and b.

   ii State the minimum value of the expression $x^2 + 12x + 11$.

b Determine the values of k for which the quadratic equation
$$x^2 + 3(k - 2)x + (k + 5) = 0$$
has equal roots. [AQA (B) Jan 2001]

11 a Solve the equation: $2x^2 + 32x + 119 = 0$.
Write your answers in the form $p + q\sqrt{2}$, where p and q are rational numbers.

b i Express $2x^2 + 32x + 119$
in the form $2(x + m)^2 + n$, where m and n are integers.

   ii Hence write down the minimum value of $2x^2 + 32x + 119$. [AQA (A) Nov 2002]

 12 a Show that the curve with equation $y = 3(x + 1)^2$ and the line with equation $y = kx - 9$ intersect when
$$3x^2 + (6 - k)x + 12 = 0.$$

b Find the values of k for which the quadratic equation
$$3x^2 + (6 - k)x + 12 = 0$$
has equal roots.

c State the geometrical relationship between the line $y = kx - 9$ and the curve $y = 3(x + 1)^2$ for these values of k. [AQA (B) Nov 2002]

13 a Express $x^2 - 6x + 7$ in the form $(x + a)^2 + b$, finding the values of a and b.

b Hence, or otherwise, find the range of values of x for which
$$x^2 - 6x + 7 < 0.$$ [AQA (A) Jan 2001]

14 Given that $f(x) = x^3 - 4x^2 - x + 4$,

a find $f(1)$ and $f(2)$,

b factorise $f(x)$ into the product of three linear factors. [AQA (A) Jan 2001]

15 It is given that:
$$f(x) = x^3 + 3x^2 - 6x - 8.$$

a Find the value of $f(2)$.

b Use the Factor Theorem to write down a factor of $f(x)$.

c Hence express $f(x)$ as a product of three linear factors. [AQA (A) Nov 2002]

16 Given that $f(x) = x^3 + 4x^2 + x - 6$:

a find $f(1)$ and $f(-1)$;

b factorise $f(x)$ into the product of three linear factors. [AQA (A) May 2002]

17 a Express $6x^3 + 17x^2 + 14x + 3$ in the form $(x + 1)(ax^2 + bx + c)$, stating the values of a, b and c.

b Hence solve the equation $6x^3 + 17x^2 + 14x + 3 = 0$. [AQA (A) Jan 2002]

18 A polynomial is given by $p(x) = 2x^3 - x^2 - 7x + 6$.

a By finding the value of $p(1)$, show that $(x - 1)$ is a factor of $p(x)$.

b Express $p(x)$ as a product of three linear factors. [AQA (B) May 2002]

19 The cubic polynomial $x^3 + ax^2 + bx + 4$, where a and b are constants, has factors $x - 2$ and $x + 1$. Use the Factor Theorem to find the values of a and b. [AQA(A) June 2001]

20 The polynomial $p(x) = x^3 + ax^2 + bx + 5$ leaves a remainder of 1 when divided by $x + 1$ and leaves a remainder of 13 when divided by $x - 2$. Find the values of the constants a and b.

21 a Express $x^2 - 8x + 13$ in the form $(x + a)^2 + b$.

b The graph of $y = x^2$ is translated to give the graph of $y = x^2 - 8x + 13$. Describe the translation.

c Sketch the graph of $y = x^2 - 8x + 13$, stating the coordinates of the vertex and the intercept with the y-axis.

22 a Express $x^2 + 2x + 3$ in the form $(x + a)^2 + b$.

b State the minimum value of $x^2 + 2x + 3$ and the value of x where this occurs.

c The graph of $y = x^2 + 2x + 3$ is translated to give the graph of $y = x^2$. Describe the translation.

23 The circle $x^2 + y^2 = 49$ has centre $(0, 0)$ and radius 7 units. It is translated to the circle $(x + 7)^2 + (y - 7)^2 = 49$.

a Describe the translation.

b Sketch the circle $(x + 7)^2 + (y - 7)^2 = 49$.

2 Coordinate geometry

2.1 The equation of a straight line

Equation of a straight line, including the forms $y - y_1 = m(x - x_1)$ and $ax + by + c = 0$.

Mid-point

The mid-point of the line segment joining $A(x_1, y_1)$ and $B(x_2, y_2)$ is given by the formula

$$\text{Mid-point} = \left(\frac{x_1 + x_2}{2}, \frac{y_1 + y_2}{2}\right)$$

> **Tip:**
> The order that the coordinates are added does not matter, but make sure you add the two x-coordinates together and the two y-coordinates together.

Example 2.1 Find the mid-point of AB where A is the point $(7, -3)$ and B is the point $(-2, -5)$.

Step 1: Substitute in the mid-point formula.

Taking $(7, -3)$ as (x_1, y_1) and $(-2, -5)$ as (x_2, y_2):

$$\text{Mid-point} = \left(\frac{x_1 + x_2}{2}, \frac{y_1 + y_2}{2}\right)$$

$$= \left(\frac{7 + (-2)}{2}, \frac{(-3) + (-5)}{2}\right)$$

$$= \left(\frac{5}{2}, -4\right)$$

> **Tip:**
> Drawing a sketch can help you to see whether your answer is reasonable.

> **Tip:**
> Be very careful with negatives.

Distance between two points

The distance AB between two points $A(x_1, y_1)$ and $B(x_2, y_2)$ is given by the formula

$$AB = \sqrt{(x_2 - x_1)^2 + (y_2 - y_1)^2}$$

> **Tip:**
> Work out the brackets before trying to square them. Remember that squares can never be negative.

Example 2.2 Find the distance between A and B where A is the point $(7, -3)$ and B is the point $(-2, -5)$.

Step 1: Substitute into the formula.
Step 2: Work out each bracket, square, add, then square root.

Taking $(7, -3)$ as (x_1, y_1) and $(-2, -5)$ as (x_2, y_2):

$$AB = \sqrt{(-2 - 7)^2 + (-5 - (-3))^2}$$

$$= \sqrt{(-9)^2 + (-2)^2}$$

$$= \sqrt{85}$$

> **Note:**
> You will not be allowed a calculator in Core 1, so if the answer is not a perfect square, leave it in surd form.

Example 2.3 P is the point $(-6, 1)$ and Q is the point $(8, -1)$. The length PQ is $k\sqrt{2}$. Find the value of k.

Step 1: Substitute into the formula.
Step 2: Work out each bracket, square, add, then square root.

Taking $(-6, 1)$ as (x_1, y_1) and $(8, -1)$ as (x_2, y_2):

$$AB = \sqrt{(8 - (-6))^2 + (-1 - 1)^2}$$

$$= \sqrt{14^2 + (-2)^2}$$

$$= \sqrt{200}$$

$$= \sqrt{100 \times 2}$$

$$= 10\sqrt{2}$$

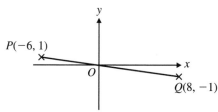

Therefore $k = 10$.

Gradient

The gradient of the line joining $A(x_1, y_1)$ and $B(x_2, y_2)$ is given by the formula

$$\text{Gradient of } AB = \frac{y_2 - y_1}{x_2 - x_1}$$

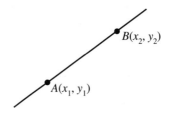

Tip:
Keep the order the same and be careful with minuses.

Example 2.4 Find the gradient of the line AB where A is the point $(7, -3)$ and B is the point $(-2, -5)$.

Step 1: Substitute into the formula.

$$\text{Gradient of } AB = \frac{-5 - (-3)}{-2 - 7} = \frac{-2}{-9} = \frac{2}{9}$$

Tip:
Substitute into the formula before attempting to work anything out.

Equations of lines

There are several ways of writing the equation of a straight line.

Gradient–intercept format
The equation of the line is $y = mx + c$, where m is the gradient and c is the y-intercept.

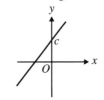

Note:
The y-intercept is the y-coordinate of the point where the line crosses the y-axis.

For example, the line $y = 3x - 2$ has gradient 3 and crosses the y-axis at $(0, -2)$.

The format $ax + by + c = 0$, where a, b and c are integers

Example 2.5 Write the equation of the straight line in the form $ax + by + c = 0$, where a, b and c are integers.

$$y = -\tfrac{1}{4}x - \tfrac{5}{6}$$

Step 1: Eliminate any fractions.
Step 2: Collect all terms on one side, with zero on the other.

$$(\times 12) \qquad 12y = -3x - 10$$
$$3x + 12y + 10 = 0$$

Tip:
To eliminate the fractions multiply *every* term by the common denominator (12).

The format $y - y_1 = m(x - x_1)$
Take P to be the general point (x, y).
A is the fixed point (x_1, y_1).

Using the gradient formula (calling the gradient m):

$$m = \frac{y - y_1}{x - x_1}$$

This can be rearranged to give

$$y - y_1 = m(x - x_1)$$

Note:
This form is useful when you know the gradient and a point on the line.

Example 2.6 Find the equation of the line with gradient $\frac{2}{3}$ that passes through the point $(-3, 1)$. Write the answer in the form $ax + by + c = 0$, where a, b and c are integers.

Step 1: Substitute into the formula.

$m = \frac{2}{3}$ and $(x_1, y_1) = (-3, 1)$.

Equation of line is

$$y - y_1 = m(x - x_1)$$

Step 2: Multiply by the denominator if necessary.

$$y - 1 = \frac{2}{3}(x - (-3))$$

$$3(y - 1) = 2(x + 3)$$

$$3y - 3 = 2x + 6$$

Step 3: Rearrange into the required format.

$$-2x + 3y - 9 = 0$$

Note:
You could write
$2x - 3y + 9 = 0$ to avoid starting with a minus sign.

Example 2.7 The points A and B have coordinates $(12, 5)$ and $(7, 3)$. Find:

a the gradient of AB

b the equation of the line AB.

Step 1: Use the gradient formula.

a Gradient $= \dfrac{3 - 5}{7 - 12} = \dfrac{-2}{-5} = \dfrac{2}{5}$

Step 2: Use the known gradient, known point formula.

b Equation of line AB is

$$y - y_1 = m(x - x_1)$$

$$y - 5 = \frac{2}{5}(x - 12)$$

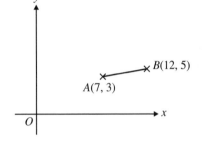

Tip:
It does not matter which point you use as (x_1, y_1).

Tip:
If you are not asked to give a specific format, then you may leave it like this.

You may have to find the **point of intersection** of two lines. In this case, solve the two equations simultaneously.

Example 2.8 Find the point of intersection of the two lines

$y = \frac{2}{3}x - 4$ and $4x - y = 9$.

Step 1: Substitute for one variable from one equation into the other.

$$y = \tfrac{2}{3}x - 4 \qquad ①$$

$$4x - y = 9 \qquad ②$$

Recall:
Simultaneous equations (Section 1.7) and points of intersection (Section 1.14).

Substitute for y from ① into ②:

$$4x - (\tfrac{2}{3}x - 4) = 9$$

Step 2: Simplify and solve the resulting equation.

$$4x - \tfrac{2}{3}x + 4 = 9$$

$$4x - \tfrac{2}{3}x = 5$$

Multiply through by 3:

$$12x - 2x = 15$$

$$x = \tfrac{3}{2}$$

Step 3: Substitute the value found into one of the original equations.

Substitute for x in ②:

$$6 - y = 9 \quad \Rightarrow \quad y = -3$$

The lines intersect at $(\tfrac{3}{2}, -3)$.

1 Write the following lines in the general form $ax + by + c = 0$.

    a $y = 3x - 2$ **b** $x + \frac{2}{3}y = 2$ **c** $\frac{1}{2}x + \frac{3}{4}y = 5$ **d** $\frac{3}{5}x = 2 - \frac{1}{2}y$

2 a Write the straight line $4x - 5y = 8$ in the form $y = mx + c$.

    b What is the gradient and point of intercept on the y-axis of the straight line $2x + 3y = 6$?

3 For the following pairs of points, calculate **i** the gradient of AB **ii** the mid-point of AB **iii** the distance AB, writing your answer in surd form if necessary.

    a $A(-3, 6), B(4, -1)$ **b** $A(4, 6), B(-2, -4)$ **c** $A(-1, -2), B(2, -1)$

4 Find the equation of the line that has:

    a a gradient of $\frac{3}{5}$ passing through the point $(-4, -2)$

    b a gradient of -3 passing through the point $(5, -3)$

    c a line passing through the points $(-3, 6)$ and $(4, -1)$. (Hint: work out the gradient first).

 5 A triangle is formed by three straight lines, $y = \frac{1}{2}x$, $2x + y + 5 = 0$ and $x + 3y - 5 = 0$.

Prove that the triangle is isosceles.

6 In the triangle ABC, A, B and C are the points $(-4, 1)$, $(-2, -3)$ and $(3, 2)$, respectively.

    a Show that ABC is isosceles.

    b Find the coordinates of the mid-point of the base.

    c Find the area of ABC.

SKILLS CHECK **2A EXTRA is on the CD**

2.2 **Parallel and perpendicular lines**

Conditions for two straight lines to be parallel or perpendicular to each other.

For two lines with gradients m_1 and m_2:

The lines are **parallel** if their gradients are the same, i.e. $m_1 = m_2$.

The lines are **perpendicular** if the product of their gradients is -1, i.e. $m_1 \times m_2 = -1$.

Note:
Perpendicular lines have gradients that are negative reciprocals of each other, i.e. $m_1 = -\dfrac{1}{m_2}$.

Example 2.9 Show that the lines $y = \frac{2}{3}x - 1$ and $2x - 3y + 6 = 0$ are parallel.

Step 1: Rearrange the equations of the lines into the form $y = mx + c$.

The first line has gradient $\frac{2}{3}$.

Rearranging the equation of the second line:

$$2x - 3y + 6 = 0$$
$$3y = 2x + 6$$
$$y = \frac{2}{3}x + 2$$

The second line has gradient $\frac{2}{3}$.

Recall:
When written in the form $y = mx + c$, the gradient is m.

Step 2: Compare gradients.   Since both lines have the same gradient, they are parallel.

Example 2.10   Find the equation of the line passing through (0, 2) that is perpendicular to the line $y = \frac{1}{4}x + 3$.

Step 1: Find the gradient of the required line.   The gradient of the given line is $\frac{1}{4}$.

The gradient of the required line is the negative reciprocal of $\frac{1}{4}$, so the gradient of the required line is -4.

Step 2: Use the gradient and known point to find the equation of the line.   The given point (0, 2) is the intercept on the y-axis, so the equation of the required line is

$y = -4x + 2.$

Tip:
Unless a particular equation is asked for, you can give any form.

Example 2.11   Find the equation of the perpendicular bisector of the line AB where A is the point $(-3, 5)$ and B is the point $(-1, 3)$.

Step 1: Find the gradient of AB.   Gradient of $AB = \dfrac{y_2 - y_1}{x_2 - x_1} = \dfrac{3 - 5}{-1 - (-3)} = \dfrac{-2}{2} = -1$

Step 2: Find the gradient of a line perpendicular to AB.   Gradient of perpendicular bisector $= 1$

Step 3: Find the mid-point of AB.   Mid-point of AB

$= \left(\dfrac{x_1 + x_2}{2}, \dfrac{y_1 + y_2}{2} \right)$

$= \left(\dfrac{(-3) + (-1)}{2}, \dfrac{5 + 3}{2} \right)$

$= (-2, 4)$

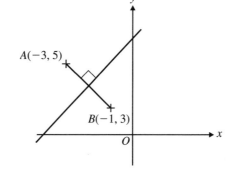

Recall:
Product of gradients is –1 for perpendicular lines.

Tip:
It may be helpful to draw a sketch.

Step 4: Find the equation of the line using $y - y_1 = m(x - x_1)$.   The perpendicular bisector has gradient 1 and passes through $(-2, 4)$. The equation of the perpendicular bisector is:

$y - 4 = 1(x - (-2))$

$y - 4 = x + 2$

$y = x + 6$

Example 2.12   The point A has coordinates $(3, -5)$ and the point B has coordinates $(5, 3)$. The mid-point of AB is M and the line MC is perpendicular to AB, where C has coordinates $(8, p)$.

a Find the coordinates of M.

b Find the gradient of MC.

c Find the value of p.

Step 1: Draw a sketch.

Step 2: Use the mid-point formula.   **a** Mid-point of $AB = \left(\dfrac{3 + 5}{2}, \dfrac{(-5) + 3}{2} \right) = (4, -1)$

Step 3: Use the gradient of the given line to find the gradient of the perpendicular line.   **b** Gradient of $AB = \left(\dfrac{3 - (-5)}{5 - 3} \right) = \dfrac{8}{2} = 4$

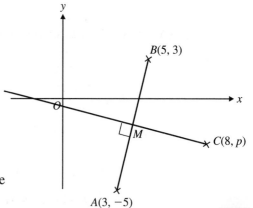

Since AB and MC are perpendicular, the product of the gradients is -1. So the gradient of MC is $-\frac{1}{4}$.

Step 4: Use the gradient formula for the given points.

c Using points $M(4, -1)$ and $C(8, p)$:

gradient of $MC = \dfrac{p - (-1)}{8 - 4} = \dfrac{p + 1}{4}$.

So $\dfrac{p + 1}{4} = -\dfrac{1}{4}$

$p + 1 = -1$

$p = -2$

Example 2.13 The line AB has equation $3x - 2y = 13$ and the points A and B have coordinates $(3, -2)$ and $(5, k)$ respectively.

  a Find the value of k.

  b Find the equation of the line through A that is perpendicular to AB, giving your answer in the form $ax + by + c = 0$, where a, b and c are integers.

Step 1: Substitute coordinates of B into equation of line.

a Since $(5, k)$ lies on the line $3x - 2y = 13$:

$3 \times 5 - 2k = 13$

$k = 1$

> **Tip:**
> If $(5, k)$ lies on the line, then $x = 5$ and $y = k$ must satisfy the equation of the line.

Step 2: Find the gradient of AB.

b A is the point $(3, -2)$ B is the point $(5, 1)$.

Gradient of $AB = \dfrac{1 - (-2)}{5 - 3} = \frac{3}{2}$

> **Tip:**
> Leave the gradient as a top-heavy fraction.

Step 3: Use the condition for gradients of perpendicular lines.

Gradient of line perpendicular to $AB = -\frac{2}{3}$

Equation of line through $A(3, -2)$ with gradient $-\frac{2}{3}$ is given by:

Step 4: Use $y - y_1 = m(x - x_1)$.

$y - (-2) = -\frac{2}{3}(x - 3)$

$y + 2 = -\frac{2}{3}(x - 3)$

Step 5: Rearrange into required format.

$3(y + 2) = -2(x - 3)$

$3y + 6 = -2x + 6$

$2x + 3y = 0$

> **Tip:**
> Take care with fractions and negatives when rearranging the equation.

SKILLS CHECK 2B: Parallel and perpendicular lines

1 State whether the following pairs of lines are parallel, perpendicular or neither:

  a $5y = 4x - 7$ 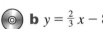 **b** $y = \frac{2}{3}x - 8$ **c** $3x - 2y + 9 = 0$

      $4y = 7 - 5x$ $4x - 6y = 5$ $2x - 3y = 6$

2 Show that the following lines are perpendicular:

  $3x - 4y = 20$ and $8x + 6y + 15 = 0$

3 Find the equation of the line parallel to the line $2x - 3y = 6$, passing through the point $(0, 3)$.

4 Find the equation of the line perpendicular to the line $y = \frac{4}{5}x - 2$, passing through the point $(0, -2)$.

5 Find the equation of the line perpendicular to $x + 5y - 10 = 0$, passing through the point $(3, -1)$.

6 Find the equation of the perpendicular bisector of AB, where $A(4, -5)$ and $B(-2, -3)$.

7 Find the equation of the line parallel to $y = -1\frac{2}{3}x - 2\frac{1}{6}$ passing through $(-2, -3)$.

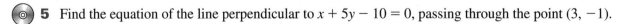

8 Find the equation of the line perpendicular to the line $y = 5x - 8$ and passing through $(2, 2)$.

 9 $A(1, 0)$, $B(3, 5)$ and $C(7, 3)$ are three vertices of a parallelogram $ABCD$. Find:

    a the gradient of AB

    b the equation of CD

    c the gradient of BC

    d the equation of AD

    e the coordinates of D.

SKILLS CHECK **2B EXTRA** is on the CD

2.3 The equation of a circle

The equation of a circle in the form $(x - a)^2 + (y - b)^2 = r^2$.

The general equation of a circle with centre (a, b) and radius r is $(x - a)^2 + (y - b)^2 = r^2$.

If the circle has centre $(0, 0)$ the equation is $x^2 + y^2 = r^2$.

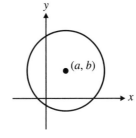

Note:
The brackets can be multiplied out to give the **expanded form** of the equation of a circle.

Example 2.14
  a State the centre and radius of the circle
    $(x + 3)^2 + (y - 4)^2 - 25 = 0$.

  b Write the equation of the circle in expanded form.

Step 1: Rearrange the equation into the general form.

  a $(x + 3)^2 + (y - 4)^2 - 25 = 0$
    $(x - (-3))^2 + (y - 4)^2 = 5^2$
    The centre is at $(-3, 4)$ and the radius is 5.

Tip:
This is a translation of $x^2 + y^2 = 25$ by $\begin{pmatrix} -3 \\ 4 \end{pmatrix}$.
See Section 1.15.

Step 2: Expand and simplify.

  b $\qquad (x + 3)^2 + (y - 4)^2 - 25 = 0$
    $x^2 + 6x + 9 + y^2 - 8y + 16 - 25 = 0$
    $\qquad\qquad x^2 + y^2 + 6x - 8y = 0$

Example 2.15 $A(3, -4)$ and $B(-5, 2)$ are the ends of a diameter of a circle. Find the equation of the circle.

Step 1: Draw a sketch.

Step 2: Find the centre of the circle.

The centre of the circle, C, is at the middle of the diameter.

$$\text{Mid-point of } AB = \left(\frac{3 + (-5)}{2}, \frac{-4 + 2}{2} \right) = (-1, -1)$$

so the centre of the circle is at $(-1, -1)$.

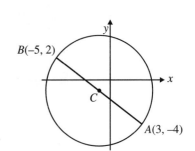

Step 3: Find the radius.

$$\text{Radius} = AC = \sqrt{(-1 - 3)^2 + (-1 - (-4))^2} = 5$$

Step 4: Write the equation of the circle.

The equation of the circle is
$$(x - (-1))^2 + (y - (-1))^2 = 5^2$$
$$(x + 1)^2 + (y + 1)^2 = 25$$

Tip:
Drawing a sketch will help you to see the situation and plan a strategy.

Completing the square

Recall:
Completing the square (Section 1.5).

You will recall from work on quadratic functions that
$x^2 + 2ax = (x + a)^2 - a^2$.

You can use this to write the expanded form of the equation of a circle in the general form.

Example 2.16 A circle C has equation $x^2 + y^2 + 6x - 8y + 18 = 0$.

    a By completing the square, express this equation in the form $(x - a)^2 + (y - b)^2 = r^2$.

    b Write down the radius and the centre of the circle.

    c Describe a geometrical transformation by which C can be obtained from the circle with equation $x^2 + y^2 = r^2$.

Step 1: Rewrite, grouping the x terms and y terms.
Step 2: Complete the square for both x and y.
Step 3: Collect all constant terms on the right-hand side.

a
$$x^2 + 6x + y^2 - 8y + 18 = 0$$
$$(x + 3)^2 - 9 + (y - 4)^2 - 16 + 18 = 0$$
$$(x + 3)^2 + (y - 4)^2 = 7$$

b The centre is at $(-3, 4)$ and the radius is $\sqrt{7}$.

Step 4: Compare with the general equation of a circle.

Step 5: Apply translation.

c Translate O, the centre of the circle $x^2 + y^2 = 7$, three units to the left and four units up to obtain C, i.e. translate by the vector $\begin{pmatrix} -3 \\ 4 \end{pmatrix}$.

Recall:
Translations (Section 1.15).

Example 2.17 Find the centre and radius of the circle $x^2 + y^2 = 10y$ and show the circle on a sketch.

Step 1: Rewrite, grouping the x terms and y terms.
$$x^2 + y^2 = 10y$$
$$x^2 + y^2 - 10y = 0$$

Step 2: Complete the square for both x and y.
$$x^2 + (y - 5)^2 - 25 = 0$$

Step 3: Collect all constant terms on the right-hand side.
$$x^2 + (y - 5)^2 = 25$$

Step 4: Sketch the circle.

Hence the centre is at $(0, 5)$ and the radius is 5.

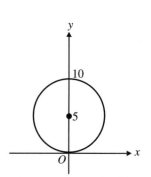

Tip:
Translate $x^2 + y^2 = 25$ by the vector $\begin{pmatrix} 0 \\ 5 \end{pmatrix}$.

SKILLS CHECK 2C: The equation of a circle

1 Give the general equation of the following circles.

    a Centre $(3, -2)$, radius 4   **b** Centre $(-5, 0)$, radius 5

2 Write down the centre and radius of the following circles.

    a $(x - 2)^2 + (y - 4)^2 = 6^2$ **b** $x^2 + (y + 3)^2 = 16$

3 Write the equation of the following circles in the expanded form.

    a $(x + 1)^2 + (y - 2)^2 = 5^2$ **b** $(x + 3)^2 + (y - 4)^2 = 25$

    c $(x - 2)^2 + (y - 6)^2 = 20$ **d** $(x + 2)^2 + (y - 2)^2 = 16$

4 Write the following circles in general form and hence state the centre and radius.

    a $x^2 + y^2 - 2x - 4y - 20 = 0$ **b** $x^2 + y^2 + 10x + 24y = 0$

    c $x^2 + y^2 - 6x + 10y + 18 = 0$ **d** $x^2 + y^2 + 2x - 6y + 3 = 0$

 5 Find the equation of the circle with centre $O(2, -5)$ and point $A(6, -8)$ on the circumference.

6 Find the equation of the circle with diameter AB where $A = (4, -5)$ and $B = (-2, -3)$.

 7 Find the centres and radii of the circles $x^2 + y^2 - 2x - 2y - 3 = 0$ and $x^2 + y^2 - 14x - 8y + 45 = 0$. Hence show that the circles touch each other.

8 The circle with equation $x^2 + y^2 - 4x - 8y + 7 = 0$ has centre C. The point $P(5, 2)$ lies on the circle.

    a Find the gradient of PC.

    b Find the equation of the line passing through P that is at right angles to the radius.

9 A circle C has equation $x^2 + y^2 = 4x - 6y + 3$.

    a Write the equation in the form $(x - a)^2 + (y - b)^2 = r^2$.

    b Write down the radius and the coordinates of the centre of the circle.

    c Describe a geometrical transformation by which C can be obtained from the circle with equation $x^2 + y^2 = r^2$.

 10 A circle with equation $x^2 + y^2 = 49$ is translated 3 units down and 5 units to the right. Write down the equation of the translated curve in expanded form.

SKILLS CHECK **2C EXTRA** is on the CD

2.4 Coordinate geometry of a circle

Coordinate geometry of a circle.

There are several circle properties that are useful in solving circle problems. Make sure that you can apply the following:

Circle property 1: angle in a semicircle

The angle subtended at the circumference of a circle by a diameter is a right angle.

Another way of saying this is: the angle in a semicircle is a right angle.

Example 2.18 **a** Show that the triangle ABC is right-angled, where A is $(0, -2)$, B is $(6, 6)$ and C is $(7, -1)$.

               b Given that A, B and C lie on the circumference of a circle, find the centre and radius of the circle and write down its equation.

> **Tip:**
> A fairly accurate sketch can be useful to help decide on a strategy to solve the problem. In this case it helps to identify where the right angle is.

Step 1: Do a sketch.   **a**   Using gradient $= \dfrac{y_2 - y_1}{x_2 - x_1}$

Step 2: Show that two sides of the triangle are perpendicular.

Gradient $BC = \dfrac{-1 - 6}{7 - 6} = -7$

Gradient $AC = \dfrac{-1 - (-2)}{7 - 0} = \dfrac{1}{7}$

Since $\text{grad}_{BC} \times \text{grad}_{AC} = -1$, BC and AC are perpendicular and the triangle ABC is right-angled at C.

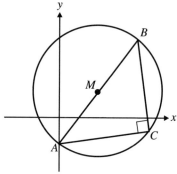

Step 3: Use circle properties to find the centre and radius.   **b**   The angle in a semicircle is a right angle, so AB is a diameter of the circle. The centre of the circle is the mid-point of AB, call this M. The radius is given by the length of AM.

Coordinates of M

$= \left(\dfrac{x_1 + x_2}{2}, \dfrac{y_1 + y_2}{2} \right)$

$= \left(\dfrac{0 + 6}{2}, \dfrac{-2 + 6}{2} \right)$

$= (3, 2)$

$AM = \sqrt{(3 - 0)^2 + (2 - (-2))^2} = 5$

The circle has centre $(3, 2)$ and radius 5.
The equation of the circle is $(x - 3)^2 + (y - 2)^2 = 25$.

Note:
You could find AB^2, AC^2 and BC^2 and check Pythagoras' Theorem.

Example 2.19 $A(-5, -1)$, $B(7, 3)$ and $C(-1, 7)$ lie on a circle with diameter AB and centre P.

    a   Find the coordinates of P.

    b   Find the equation of the perpendicular bisector of AC, writing your answer in the form $ax + by + c = 0$, where a, b and c are integers.

    c   Show that the perpendicular bisector of AC passes through P.

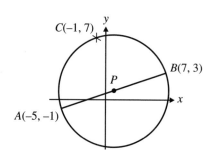

Step 1: Find the centre of the circle using circle properties.   **a** AB is a diameter, so P is at the mid-point of AB.

Coordinates of $P = \left(\dfrac{-5 + 7}{2}, \dfrac{-1 + 3}{2} \right) = (1, 1)$

Step 2: Put the additional information on the sketch.   **b**   Gradient of $AC = \dfrac{7 - (-1)}{-1 - (-5)} = \dfrac{8}{4} = 2$

Gradient of perpendicular bisector $= -\dfrac{1}{2}$

Step 3: Find the equation of the perpendicular bisector of AC.   Mid-point of $AC = \left(\dfrac{-5 + (-1)}{2}, \dfrac{-1 + 7}{2} \right) = (-3, 3)$

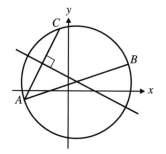

Recall:
The product of the gradients of perpendicular lines is -1 (Section 2.2).

The perpendicular bisector has gradient $-\frac{1}{2}$ and goes through $(-3, 3)$. The equation of the perpendicular bisector of AC is

$$y - 3 = -\tfrac{1}{2}(x - (-3))$$
$$2(y - 3) = -(x + 3)$$
$$2y - 6 = -x - 3$$
$$x + 2y - 3 = 0$$

Recall:
The equation of the line through (x_1, y_1) with gradient m is $y - y_1 = m(x - x_1)$ (Section 2.1).

Step 4: Check that the coordinates of P satisfy the equation of the line.

c When $x = 1$ and $y = 1$:

$$x + 2y - 3 = 1 + 2 - 3 = 0$$

Therefore $(1, 1)$ lies on the line $x + 2y - 3 = 0$ so the perpendicular bisector of AC passes through P.

Circle property 2: perpendicular to a chord

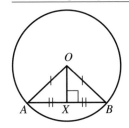

The perpendicular from the centre to a chord bisects the chord.

Note also:

- OAB is an isosceles triangle
- OX is the perpendicular bisector of chord AB.

Example 2.20 AB is a chord of a circle. A is the point $(-2, 1)$ and B is the point $(4, -1)$. The centre of the circle is $C(2, k)$.

a Find the mid-point M of AB.

b Find the gradient of AB.

c Find the equation of the line MC.

d Hence, or otherwise, find the value of k.

Step 1: Use the mid-point formula.

a $M = \left(\dfrac{-2 + 4}{2}, \dfrac{1 + (-1)}{2}\right) = (1, 0)$

Tip:
Show M on the sketch.

Step 2: Use the gradient formula.

b Gradient of $AB = \dfrac{-1 - 1}{4 - (-2)} = -\dfrac{1}{3}$

Step 3: Find the equation of the perpendicular bisector of AB.

c $AM = MB \Rightarrow CM$ is the perpendicular bisector of AB.

Gradient of $CM = 3$

Equation of CM:
$$y - 0 = 3(x - 1)$$
$$y = 3x - 3$$

Tip:
Use $m_1 \times m_2 = -1$ for perpendicular lines.

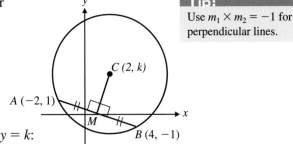

Step 4: Substitute point into the equation of the line.

d C lies on the line, so when $x = 2$ and $y = k$:

$$k = 3 \times 2 - 3 = 3$$

So $k = 3$.

Example 2.21 A circle with centre $C(1, 4)$ has chord AB where A is the point $(-4, 2)$ and B is the point $(3, k)$. Find the possible values of k.

Note:
At this stage you don't know exactly where B is. But you do know it lies on $x = 3$, so there are two possible positions.

Step 1: Draw a sketch.

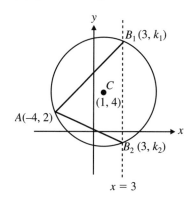

Step 2: Calculate the squares of the distances from C to A and B.

$CA^2 = (1 - (-4))^2 + (4 - 2)^2 = 29$

$CB^2 = (1 - 3)^2 + (4 - k)^2 = 4 + 16 - 8k + k^2 = 20 - 8k + k^2$

Step 3: Equate the distances and solve for k.

$CB = CA$ (both are radii) $\Rightarrow CB^2 = CA^2$

$20 - 8k + k^2 = 29$

$k^2 - 8k - 9 = 0$

$(k - 9)(k + 1) = 0$

$k = 9$ or $k = -1$

Recall:
Quadratic equations (Section 1.6).

Note:
The two values for k confirm the two possible positions for B of $(3, 9)$ and $(3, -1)$.

Circle property 3: tangent to a circle

The tangent to a circle is perpendicular to the radius at its point of contact.

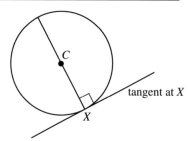

Example 2.22 $A(-1, 6)$ is a point on the circumference of a circle, centre $C(5, -3)$.
Find the gradient of the tangent at A.

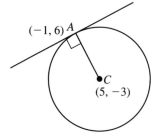

Step 1: Draw a sketch.
Step 2: Find the gradient of the radius at A.

Gradient $CA = \dfrac{-3 - 6}{5 - (-1)} = \dfrac{-9}{6} = -\dfrac{3}{2}$

Step 3: Find the gradient of the line perpendicular to the radius.

The tangent is perpendicular to the radius $\Rightarrow$ the gradient of the tangent is $\frac{2}{3}$.

Tip:
Leave gradients as top heavy fractions.

Example 2.23 A circle has centre $C(-3, 2)$.
The line $y = x - 1$ is a tangent to the circle at A.

a Find the gradient of CA.

b Find the equation of CA.

c Find the coordinates of A.

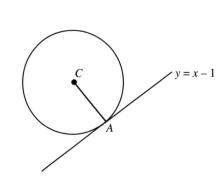

Step 1: Use circle properties.

a Gradient of tangent $= 1$

The radius is perpendicular to the tangent

$\Rightarrow$ gradient of $CA = -1$

Recall:
The gradient of the line $y = mx + c$ is m (Section 2.1).

Step 2: Use the equation of a line $y - y_1 = m(x - x_1)$.

b Equation of CA:

$y - 2 = -1(x - (-3))$

$y - 2 = -(x + 3)$

$y - 2 = -x - 3$

$\quad y = -x - 1$

Recall:
Product of gradients of perpendicular lines is -1 (Section 2.2).

Step 3: Find the point of intersection of the lines by solving the simultaneous equations.

c A is the point of intersection of the lines

$y = x - 1 \qquad \textcircled{1}$

$y = -x - 1 \qquad \textcircled{2}$

Substituting for y from $\textcircled{1}$ into $\textcircled{2}$

$x - 1 = -x - 1$

$\quad 2x = 0$

$\quad\ x = 0$

Substituting for x in $\textcircled{1}$

$y = 0 - 1 = -1$

So the coordinates of A are $(0, -1)$.

Recall:
Solving two linear simultaneous equations (Section 1.7).

SKILLS CHECK **2D: Properties of circles**

 1 $A(-4, 5)$, $B(4, 3)$ and $P(1, 0)$ are the vertices of triangle ABP.

   a Show that angle $APB = 90°$.

   b Hence show that P lies on the circumference of a circle with diameter AB.

   c The circle has centre C. Find the coordinates of C and the radius of the circle.

   d Hence write down the equation of the circle.

2 The point $(2, k)$ lies on the circumference of a circle with diameter AB, where A is the point $(-4, 1)$ and B is the point $(3, 2)$. Find the possible values of k.

3 AB is the diameter of a circle, where $A(2, 2)$ and $B(p, q)$ lie on the line $x + 2y - 6 = 0$. $C(8, 2)$ lies on the circumference of the circle.

   a Draw a sketch showing the line, A and C.

   b Find the equation of the line CB.

   c Find the values of p and q.

Tip:
Hint to **b**: AB is a diameter, C is on the circumference. What size is angle ACB?

 4 $A(1, -2)$ and $B(-5, 4)$ are the ends of a chord of a circle and C is the point $(-1, 2)$.

   a Show that triangle ACB is isosceles.

   b Could C be the centre of the circle? Give a reason for your answer.

5 $A(-3, 3)$ and $B(5, 1)$ are the two ends of a chord AB of a circle.

   a Find the mid-point of AB.

   b Show, by using gradients, that $C(0, -2)$ could be the centre of the circle.

6 AB and CD are two chords of a circle, where A is $(1, 5)$, B is $(5, 3)$, C is $(3, -1)$ and D is $(5, 1)$.

    a Find the equation of the perpendicular bisector of AB.

    b Find the equation of the perpeneicular bisector of CD.

    c Hence find the coordinates of the centre of the circle.

7 $M(1, -3)$ is the mid-point of a chord AB of the circle $x^2 + y^2 = 20$. Find the equation of AB. CD

8 The point $P(5, 1)$ lies on a circle with centre $C(1, 4)$.

    a Show that P also lies on the line L with equation $3y - 4x + 17 = 0$.

    b Show that CP is perpendicular to L.

    c Deduce that L is a tangent to the circle at P.

SKILLS CHECK **2D EXTRA is on the CD**

2.5 Tangents and normals to circles

The equation of the tangent and normal at a given point on the curve.

A tangent and a normal can be drawn at any point on the circumference of a circle.

Since the radius is perpendicular to a tangent, the normal will pass through the centre of the circle.

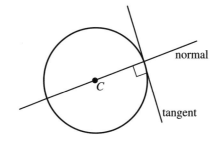

Note:
A normal is the line perpendicular to a tangent.

Example 2.24 Find the equation of the tangent to the circle, centre $C(4, 3)$, at the point $A(5, -2)$ on the circumference. Write your answer in the form $ax + by + c = 0$, where a, b and c are integers.

Step 1: Draw a sketch.

Step 2: Find the gradient of the radius and use it to find the gradient of the tangent.

Gradient $CA = \dfrac{-2 - 3}{5 - 4} = -5$

Gradient of tangent $= \frac{1}{5}$

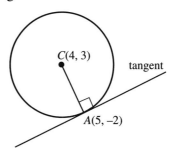

Step 3: Use $y - y_1 = m(x - x_1)$ to find the equation of the tangent.

Equation of tangent at A:

$$y - (-2) = \tfrac{1}{5}(x - 5)$$
$$5(y + 2) = x - 5$$
$$5y + 10 = x - 5$$
$$x - 5y - 15 = 0$$

Example 2.25 The point $(6, 8)$ is on the circumference of a circle with centre $C(2, 5)$. Find the equation of the normal to the circle at A.

Step 1: Draw a sketch.

Step 2: Find the gradient of the radius.

The normal at A passes through C.

Gradient $CA = \dfrac{8 - 5}{6 - 2} = \dfrac{3}{4}$

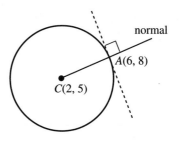

Step 3: Use $y - y_1 = m(x - x_1)$ to find the equation of the normal.

Equation of CA:

$$y - 8 = \tfrac{3}{4}(x - 6)$$

Tip:
You can use $C(2, 5)$ or $A(6, 8)$ as (x_1, y_1) when finding the equation of the line.

Tip:
If the form of the equation is not stated, it may be left unsimplified.

1 Find the equation of the tangent to the circle, centre $C(3, 5)$, at the point $A(1, 3)$.

2 Find the equation of the tangent to the circle $x^2 + y^2 + 4x - 6y + 8 = 0$ at the point $A(-1, 1)$.

3 Find the equation of the normal to the circle, centre $C(2, 2)$, at the point $A(5, 1)$.

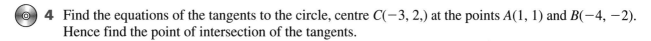 **4** Find the equations of the tangents to the circle, centre $C(-3, 2,)$ at the points $A(1, 1)$ and $B(-4, -2)$. Hence find the point of intersection of the tangents.

5 $A(-8, -2)$ and $B(2, -2)$ are points on the circumference of a circle with centre $O(-3, \frac{1}{2})$. Find the equation of the normal at A and the equation of the tangent at B.

6 **a** Find the centre and radius of the circle $x^2 + y^2 - 4x + 10y + 4 = 0$ and hence sketch the circle.

   b Show that $P(6, -2)$ lies on the circle.

   c Find the equation of the normal to the circle at P, giving your answer in the form $ax + by + c = 0$, where a, b and c are integers.

7 $A(5, -1)$ and $B(7, 5)$ are two ends of a chord AB of a circle, centre $O(3, 3)$. Show that the tangents at A and B are perpendicular.

2.6 The intersection of a straight line and a curve

The intersection of a straight line and a curve.

This section is concerned with the intersection of a straight line with a circle or with a graph of a quadratic function.

When finding the point(s) of intersection, you may have to solve a quadratic equation $ax^2 + bx + c = 0$. Here are the conditions relating to the discriminant of the quadratic polynomial $ax^2 + bx + c$. There are three cases to consider:

> **Recall:**
> The discriminant of a quadratic function (Section 1.3).

1 If $b^2 - 4ac > 0$ there are two real roots. This means there are two distinct points of intersection and the line cuts the curve twice.

2 If $b^2 - 4ac = 0$ there is one real root, so there is one point of intersection. The line touches the curve, so the line is a tangent to the curve at the point of intersection.

3 If $b^2 - 4ac < 0$ there are no real roots. The line and curve do not intersect.

Example 2.26 Find the points of intersection of the circle $x^2 + y^2 = 20$ and the line $y = 3x - 10$.

Step 1: Substitute for y into the equation of the circle.

At the points of intersection, both these equations are satisfied.
$$y = 3x - 10 \qquad ①$$
$$x^2 + y^2 = 20 \qquad ②$$

Substituting for y from ① into ②:
$$x^2 + (3x - 10)^2 = 20$$

Step 2: Expand and rearrange into $f(x) = 0$.
$$x^2 + 9x^2 - 60x + 100 = 20$$
$$10x^2 - 60x + 80 = 0$$

Step 3: Solve the quadratic equation.
$$(\div 10) \quad x^2 - 6x + 8 = 0$$
$$(x - 4)(x - 2) = 0$$
$$x = 4 \text{ or } x = 2$$

Step 4: Find the corresponding y-coordinate.

Substituting in ①:
when $x = 4$, $y = 3 \times 4 - 10 = 2$
when $x = 2$, $y = 3 \times 2 - 10 = -4$
The line and circle intersect at $(4, 2)$ and $(2, -4)$.

> **Recall:**
> Solving simultaneous equations (Section 1.7).

> **Note:**
> There may be two answers and they will be coordinates: there will be an x-value and a y-value.

> **Note:**
> For the quadratic equation $x^2 - 6x + 8 = 0$, $a = 1$, $b = -6$, $c = 8$.
> The discriminant:
> $b^2 - 4ac = 36 - 32 > 0 \Rightarrow$ there are two points of intersection.

Example 2.27 Show that the line $y - x = -8$ and the circle $x^2 + y^2 - 6x - 8y = 0$ do not intersect.

Step 1: Rearrange the linear equation to make x or y the subject.
$$y - x = -8 \quad \Rightarrow y = x - 8$$

Step 2: Substitute for y into the equation of the circle.

Substituting for y in the equation of the circle:
$$x^2 + (x - 8)^2 - 6x - 8(x - 8) = 0$$

Step 3: Expand and rearrange into $f(x) = 0$.
$$x^2 + x^2 - 16x + 64 - 6x - 8x + 64 = 0$$
$$2x^2 - 30x + 128 = 0$$
$$(\div 2) \qquad x^2 - 15x + 64 = 0$$

$$a = 1, b = -15, c = 64$$

Step 4: Test the discriminant.
$$b^2 - 4ac = 15^2 - 4 \times 1 \times 64 = 225 - 256 = -31$$

Since $b^2 - 4ac < 0$ the equation has no real roots.
Therefore the line and the circle do not intersect.

Example 2.28 The line $y = 4x + k$ is a tangent to the curve $y = x^2 + 2x$. Find k.

Step 1: Substitute for y into the equation of the circle.

The line and curve touch when both these equations are satisfied.

$$y = 4x + k \qquad \text{①}$$
$$y = x^2 + 2x \qquad \text{②}$$

Substituting for y from ① into ②

Step 2: Expand and rearrange into $f(x) = 0$.

$$4x + k = x^2 + 2x$$
$$x^2 - 2x - k = 0$$

Step 3: Use the condition for the discriminant.

Since the line is a tangent, $x^2 - 2x - k = 0$ has one root.
$$a = 1, b = -2, c = -k$$
$$b^2 - 4ac = 0 \Rightarrow (-2)^2 - 4 \times 1 \times (-k) = 0$$
$$4 + 4k = 0$$
$$k = -1$$

SKILLS CHECK **2F: Intersection of lines and curves**

1 Find the point(s) of intersection of the following lines and curves.

  a $x^2 + y^2 = 25$ and $x + y = 7$

  b $y = x^2$ and $y = 4x - 4$

  c $x^2 + y^2 - 10x - 4y + 12 = 0$ and $y = x + 2$

  d $x^2 + y^2 - 4x - 2y - 15 = 0$ and $2x + y + 5 = 0$

2 Show that the line $y = x - 3$ is a tangent to the circle with centre $(-2, 1)$, radius $\sqrt{18}$, and find the point of contact.

 3 Show that if the line $y = 2x - k$ does not intersect the curve $y = x^2 - 2$ then $k > 3$.

4 Solve the following pairs of simultaneous equations, if possible, and interpret each result geometrically.

  a $y = x^2 + 2x - 3$ and $y = 4x - 4$ **b** $(x - 3)^2 + (y + 4)^2 = 25$ and $y = 0$

  c $y = 2x^2$ and $y = 8x - 8$ **d** $x^2 + y^2 = 2x$ and $x + y = 3$

5 The line $y = x + k$ is a tangent to the curve $y = x^2 - x + 2$. Find the value of k.

 6 The straight line $y = x + 7$ intersects the circle $x^2 + y^2 = 25$ at the points P and Q.

  a Show that the x-coordinates of P and Q satisfy the equation $x^2 + 7x + 12 = 0$.

  b Hence find the coordinates of P and Q.

 7 A circle, centre $C(4, 3)$ touches the line with equation $3x - 2y + 7 = 0$ at P.

  a Find the equation of CP.

  b By solving the equations simultaneously, find the coordinates of P.

  c Find the equation of the circle.

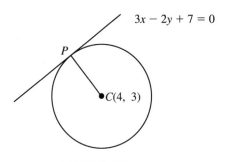

SKILLS CHECK **2F EXTRA is on the CD**

1 The point A has coordinates $(2, 3)$ and O is the origin.

   a Write down the gradient of OA and hence find the equation of the line OA.

   b Show that the line which has equation $4x + 6y = 13$:
      i   is perpendicular to OA;
      ii passes through the midpoint of OA.   [AQA (A) May 2003]

 2 A circle with centre C has diameter AB, where A is the point $(3, 6)$ and B is $(7, -4)$.

   a Find the coordinates of C and the radius of the circle. Hence, or otherwise, find a cartesian equation for the circle.

   b The point P has coordinates $(1, -2)$.
      Prove that the point P lies inside the circle.   [AQA (B) June 2002]

3 The line $2x - y + 6 = 0$ intersects the coordinate axes at two points $A(a, 0)$ and $B(0, b)$.

   a Find the values of a and b.

   b Find the coordinates of M, the mid-point of AB.

   c Find the equation of the line through M perpendicular to AB, giving your answer in the form $y = mx + c$.   [AQA (A) Jan 2003]

 4 The line AB has equation $5x - 2y = 7$, and the point A has coordinates $(1, -1)$ and the point B has coordinates $(3, k)$.

   a i   Find the value of k.
      ii Find the gradient of AB.

   b Find an equation for the line through A which is perpendicular to AB.

   c The point C has coordinates $(-6, -2)$. Show that AC has length $p\sqrt{2}$, stating the value of p.   [AQA (B) Jan 2003]

5 A circle has the equation $x^2 + y^2 + 4x - 14y + 4 = 0$.

   a Find the radius of the circle and the coordinates of its centre.

   b Sketch the circle.   [AQA (A) June 2003]

6 The points A, B and C have coordinates $(1, 7)$, $(5, 5)$ and $(7, 9)$ respectively.

   a Show that AB and BC are perpendicular.

   b Find an equation of the line BC.

   c The equation of the line AC is $3y = x + 20$ and M is the mid-point of AB.
      i   Find an equation of the line through M parallel to AC.
      ii This line intersects BC at the point T. Find the coordinates of T.   [AQA (B) June 2001]

7 The equation of the line AB is $5x - 3y = 26$.

   a Find the gradient of AB.

   b The point A has coordinates $(4, -2)$ and a point C has coordinates $(-6, 4)$.

     i Prove that AC is perpendicular to AB.

     ii Find an equation for the line AC, expressing your answer in the form $px + qy = r$, where p, q and r are integers.

   c The line with equation $x + 2y = 13$ also passes through the point B.
Find the coordinates of B.   [AQA (B) June 2002]

8 The points P, Q and R have coordinates $(3, 1)$, $(1, 9)$ and $(7, 2)$ respectively.

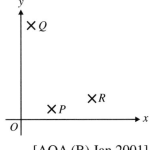

   a Find an equation for the straight line QR in the form $ax + by = c$, where a, b and c are integers.

   b Prove that the triangle PQR is right-angled and find its area.

   c Determine an equation for the straight line which passes through P and which is perpendicular to QR.

                                     [AQA (B) Jan 2001]

9 The points A and B have coordinates $(1, 7)$ and $(15, 5)$ respectively. The point O is the origin.

   a Find an equation for the straight line AB in the form $ax + by = c$, where a, b and c are integers.

   b Show that the lines OA and AB are perpendicular.

   c A circle passes through the points O, A and B.

     i Explain briefly why OB is a diameter.

     ii Write down the coordinates of the centre of the circle.

     iii Find the radius of the circle in the form $p\sqrt{10}$, where p is a rational number.

                                     [AQA (B) Jan 2002]

10 The line joining the points $A(0, 5)$ and $B(4, 1)$ is a tangent to a circle whose centre, C, is at the point $(5, 4)$.

   a Find the equation of the line AB.

   b Find the equation of the line through C which is perpendicular to AB.

   c Find the coordinates of the point of contact of the line AB with the circle.

   d Find the equation of the circle.   [AQA (A) Jan 2002]

11 The point A has coordinates $(3, -5)$ and the point B has coordinates $(1, 1)$.

   a **i** Find the gradient of AB.

     ii Show that the equation of the line AB can be written in the form $rx + y = s$, where r and s are positive integers.

   b The mid-point of AB is M and the line MC is perpendicular to AB.

     i Write down the coordinates of M.

     ii Find the gradient of the line MC.

     iii Given that C has coordinates $(5, p)$, find the value of the constant p.

12 The equation of a circle is $x^2 + y^2 - 2x + 6y - 15 = 0$.

    a Find the centre and radius of the circle.

    b The line $y = x + 1$ intersects the circle at A and B. Find the exact length of AB.

 13 **a** Solve the simultaneous equations $y = x^2 - 3x + 2$ and $y = 3x - 7$.

    b Interpret the solution to **a** geometrically.

14

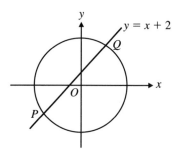

The straight line $y = x + 2$ intersects the circle $x^2 + y^2 = 16$ at the points P and Q.

    a Show that the x-coordinates of P and Q satisfy the equation $x^2 + 2x - 6 = 0$.

    b Hence find the coordinates of P and Q, giving each answer in the form $a \pm \sqrt{b}$, where a and b are integers.

15 A circle C has equation $x^2 + (y - 5)^2 = 9$ and a straight line L has equation $y = \frac{4}{3}x$.

    a Write down:

        i the coordinates of the centre of the circle C,

        ii the radius of the circle C.

    b Show that the x-coordinate of a point common to C and L must satisfy the quadratic equation $25x^2 - 120x + 144 = 0$.

    c Solve this equation for x.

    d Comment on the geometrical relationship between C and L which is shown by your answer to **c**.

3 Differentiation

3.1 Differentiation, gradients and rates of change

The derivative of f(x) as the gradient of the tangent to the graph of $y = f(x)$ at a point; the gradient of the tangent as a limit; interpretation as a rate of change.

On the curve $y = f(x)$, the slope may vary for different values of x.

At a point P on the curve, a measure of the slope is the **gradient** at P. This tells you how y is changing in relation to x at that point. This is described as the **rate of change** of y with respect to x.

Note:
The tangent at P *touches* the curve at P.

The **gradient of the curve** at P is defined as the **gradient of the tangent** at P. The tangent is the limiting position of the line through P and point Q on the curve as Q gets closer and closer to P.

The gradient of the tangent at a point can be found by **differentiating** $y = f(x)$.

Tip:
See Tangents and normals (Section 3.4).

This gives the **gradient function**, also known as the **derived function** or the **derivative** with respect to x.

The gradient function is usually written $\dfrac{dy}{dx}$ or $f'(x)$.

Note:
$\dfrac{dy}{dx}$ is read as 'dee y by dee x' and $f'(x)$ is read as 'f dashed x'.

3.2 Differentiation of polynomials

Differentiation of polynomials.

In Core 1 you need to be able to differentiate positive powers of x, using the rule:

$$y = x^n \Rightarrow \frac{dy}{dx} = nx^{n-1}$$

If a is a constant:

$$y = ax^n \Rightarrow \frac{dy}{dx} = nax^{n-1}$$

It is useful to remember the following:

$$y = ax \Rightarrow \frac{dy}{dx} = a$$

$$y = a \Rightarrow \frac{dy}{dx} = 0$$

Tip:
Multiply by the power of x and decrease the power by 1.

Note:
In function notation, $f(x) = ax^n$ $\Rightarrow f'(x) = nax^{n-1}$.

Note:
$y = ax$ is a line, with constant gradient.
$y = a$ is a line parallel to the x-axis, with zero gradient.

Example 3.1 Find $\dfrac{dy}{dx}$ when   **a** $y = x^3$ **b** $y = 4x^2$ **c** $y = 5$.

Step 1: Use the differentiation rule.

a $y = x^3$

$\dfrac{dy}{dx} = 3x^2$

b $y = 4x^2$

$\dfrac{dy}{dx} = 2 \times 4x^1 = 8x$

c $y = 5$

$\dfrac{dy}{dx} = 0$

To differentiate a **polynomial** in x containing several terms, for example $3x^7 - 2x^3 + 3x - 1$, differentiate the terms individually, using the rule:

$$y = f(x) \pm g(x) \Rightarrow \frac{dy}{dx} = f'(x) \pm g'(x)$$

Recall:
Polynomials (Section 1.9).

Example 3.2 It is given that $y = 3x^7 - 2x^3 + 3x - 1$. Find the derivative of y with respect to x.

Step 1: Differentiate each term separately.

$$y = 3x^7 - 2x^3 + 3x - 1$$
$$\frac{dy}{dx} = 21x^6 - 6x^2 + 3$$

Example 3.3 Find $f'(x)$ where $f(x) = (2x - 3)^2$.

Step 1: Expand. $\quad f(x) = (2x - 3)^2 = 4x^2 - 12x + 9$

Step 2: Differentiate term by term. $\quad f'(x) = 8x - 12$

Tip:
Write terms in index form before differentiating.

3.3 Differentiation and gradients

Applications of differentiation to gradients.

To find the gradient at a point on a curve, substitute the x-value of the point into the gradient function.

Example 3.4 The sketch shows the curve $y = (x - 3)^2$. Using differentiation, find the gradient when

a $x = 3$,

b $x = 4$.

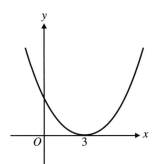

Tip:
Notice that the tangent at $x = 3$ is horizontal, so you expect the gradient to be zero. Differentiating provides a check.

Step 1: Form an expanded polynomial.

$$y = (x - 3)^2$$
$$= x^2 - 6x + 9$$

Step 2: Differentiate term by term.

$$\frac{dy}{dx} = 2x - 6$$

Tip:
Each term must be in index form before differentiating.

Step 3: Substitute the x-value into the derivative.

a At $x = 3$, $\dfrac{dy}{dx} = 2(3) - 6 = 0$

The gradient when $x = 3$ is 0.

b At $x = 4$, $\dfrac{dy}{dx} = 2(4) - 6 = 2$

The gradient when $x = 4$ is 2.

SKILLS CHECK **3A: Differentiating polynomials and finding gradients**

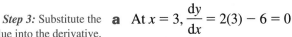

1 Differentiate with respect to x.

a $y = x^4$ **b** $f(x) = 7x^3 - 2x^2 + 3$ **c** $f(x) = (4x + 1)^2$ **d** $y = \frac{1}{2}x(2x - 1)$

 2 Find the derivative, with respect to x, of

a $y = x(x - 1)^2$ **b** $y = \dfrac{3x^4 + 5x}{2}$ **c** $y = 2$.

3 For the curve in **2a**, find the gradient of the tangent at $x = 5$.

4 For the curve in **2b**, find the gradient of the tangent at $x = -1$.

 5 **a** Sketch the graph of $y = f(x)$, where $f(x) = 3x^2 + 1$.

    b Find $f'(x)$ and use it to find the gradient of the tangent at $x = -2$. Sketch the tangent on the graph.

6 A curve has equation $y = x^2 + 5x - 4$.

    a Find $\dfrac{dy}{dx}$.

    b Determine the coordinates of the point on the curve where the gradient is 3.

7 The variables P and x are connected by the formula $P = (4x - 3)(x + 5)$. Find $\dfrac{dP}{dx}$.

8 Given that $f(x) = x(x - 2)(x + 2)$, find $f'(x)$.

9 Given that $A = 5y - 3y^2$, find $\dfrac{dA}{dy}$.

 10 A curve has equation $y = x^2(x - 3)$.

    a Find $\dfrac{dy}{dx}$.

    b Determine the point on the curve where the gradient is -3.

SKILLS CHECK **3A EXTRA** is on the CD

3.4 Tangents and normals

Applications of differentiation to tangents and normals.

The **tangent** at P is the line that touches the curve at P.

The **normal** at P is the line through P perpendicular to the tangent.

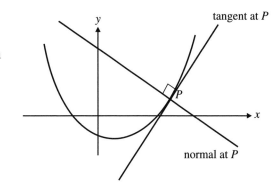

The tangent and normal are perpendicular, so the product of their gradients is -1.

To find the **equation of the tangent** to the curve $y = f(x)$ at the point $P(x_1, y_1)$, find the gradient of the tangent, then substitute into the equation of a line:

$y - y_1 = m(x - x_1)$.

To find the **equation of the normal**, first find the gradient of the tangent and use it to calculate the gradient of the normal. You can then substitute into the equation of a line, as before.

Note:
If gradient of tangent $= 2$, then gradient of normal $= -\frac{1}{2}$ (Section 2.2).

Recall:
Equation of a straight line (Section 2.1)

Note:
You could use $y = mx + c$.

53

Example 3.5 A curve has equation $y = x^3 - 6x + 1$. The point P(2, −3) lies on the curve. Giving your answers in the form $ax + by + c = 0$, where a, b and c are integers, find the equation of

    a the tangent at P

    b the normal at P.

Step 1: Find the gradient function.

  a $y = x^3 - 6x + 1$

$$\frac{dy}{dx} = 3x^2 - 6$$

Step 2: Find the gradient of the tangent at P.

When $x = 2$, $\dfrac{dy}{dx} = 3(2)^2 - 6 = 6$

The gradient of the tangent at (2, −3) is 6.

Equation of tangent at P:

Step 3: Use $y - y_1 = m(x - x_1)$.

$$y - (-3) = 6(x - 2)$$
$$y + 3 = 6x - 12$$
$$6x - y - 15 = 0$$

Note:
$(x_1, y_1) = (2, -3)$ and $m = 6$.

Step 4: Find the gradient the of normal.

  b Since product of gradients $= -1$, gradient of normal $= -\frac{1}{6}$.

Equation of normal:

Step 5: Use $y - y_1 = m(x - x_1)$.

$$y - (-3) = -\frac{1}{6}(x - 2)$$
$$y + 3 = -\frac{1}{6}(x - 2)$$
$$6(y + 3) = -x + 2$$
$$x + 6y + 16 = 0$$

Note:
$(x_1, y_1) = (2, -3)$ and $m = -\frac{1}{6}$.

3.5 Increasing and decreasing functions

Applications of differentiation to increasing and decreasing functions.

Consider a function f(x).

If, as the x-value increases, the corresponding value of f(x) increases, the function is an **increasing function**.

The gradient of the curve $y = $ f(x) is positive.

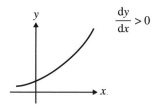

$\dfrac{dy}{dx} > 0$

Note:
Curves can be both increasing and decreasing, depending on which x-values you are considering.

If, as the x-value increases, the corresponding value of f(x) decreases, the function is a **decreasing function**.

The gradient of the curve $y = $ f(x) is negative.

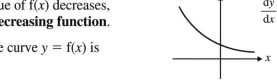

$\dfrac{dy}{dx} < 0$

Example 3.6 Find the set of values of x for which the curve $y = (x - 3)^2$ is increasing.

Step 1: Find $\dfrac{dy}{dx}$.

$$y = (x - 3)^2$$
$$= x^2 - 6x + 9$$
$$\frac{dy}{dx} = 2x - 6$$

Step 2: Set $\dfrac{dy}{dx} > 0$.

For an increasing function, $\dfrac{dy}{dx} > 0$

$$\Rightarrow 2x - 6 > 0$$

Step 3: Solve the inequality.

$$x > 3$$

Recall:
Inequalities (Section 1.8).

Note:
This agrees with the sketch in Example 3.4.

Example 3.7 Find the set of values for which the function $f(x) = 4x^2 - x$ is decreasing.

Step 1: Find $f'(x)$. $f(x) = 4x^2 - x \Rightarrow f'(x) = 8x - 1$

Step 2: Set $f'(x) < 0$.   For a decreasing function, $f'(x) < 0$
$$\Rightarrow 8x - 1 < 0$$

Step 3: Solve the inequality. $x < \frac{1}{8}$

SKILLS CHECK **3B: Applications of differentiation**

1 Find the equation of the tangent to the curve $y = x^3 - 3$ at the point $(1, -2)$.

2 A curve has equation $y = x^3 + 2x - 1$.

    a The curve goes through the point $P(1, q)$. Find q.

    b Find the gradient at P.

    c Find the equation of the tangent at P, giving your answer in the form $y = mx + c$.

    d Find the equation of the normal at P, giving your answer in the form $ax + by + c = 0$.

3 Find the set of values of x for which the function is **i** increasing **ii** decreasing:

    a $f(x) = (x - 2)^2$ **b** $f(x) = x^2 + 8$ **c** $g(x) = x^3 - 3x + 2$

4 The sketch shows the curve $y = x^2 - 2x + 3$ and the normal to the curve at $(0, 3)$.

    The normal intersects the curve again at A.

    a Find the gradient of the tangent at $(0, 3)$.

    b Find the equation of the normal at $(0, 3)$.

    c Find the x-coordinate of A.

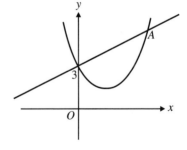

 5 **a** Find the equation of the normal to the curve $y = x^2 + 3$ at the point where $x = -1$.

    b By solving simultaneous equations, find the coordinates of the point where the normal meets the curve again.

    c Draw a sketch to illustrate the information.

6 A curve has equation $y = (x - 2)^2 - 3$.

    a Find the equation of the tangent at the point on the curve where $x = -1$.

    b Find the coordinates of the point on the curve where the normal is parallel to the line $y = \frac{1}{2}x + 5$.

7 It is given that $f(x) = (2x - 1)(3x + 1)$. Find the values of x for which $f(x)$ is an increasing function.

8 A curve has equation $y = x^3 - 8x + 4$. The point P has x-coordinate 2.

    a Find the equation of the tangent to the curve at P.

    b The tangent crosses the x-axis at A and the y-axis at B. Find the coordinates of A and B.

    c Find the area of triangle OAB, where O is the origin.

SKILLS CHECK **3B EXTRA** is on the CD

3.6 Stationary points

Applications of differentiation to maxima, minima and stationary points.

Second order derivatives.

At a **stationary point** on a curve, the gradient is zero, so $\dfrac{dy}{dx} = 0$.

In Core 1, the stationary points studied are either maximum turning points or minimum turning points.

Note:
The tangent at a stationary point is horizontal.

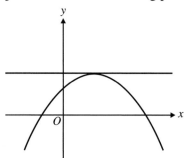

Maximum point
(the turn is at the top)

Minimum point
(the turn is at the bottom)

Example 3.8 By differentiating, find the coordinates of the stationary point of the curve $y = (x - 2)^2 + 3$.

Step 1: Find $\dfrac{dy}{dx}$.

$$y = (x - 2)^2 + 3$$
$$= x^2 - 4x + 7$$
$$\frac{dy}{dx} = 2x - 4$$

Recall:
Expand and simplify first.

Step 2: Put $\dfrac{dy}{dx} = 0$ and solve for x.

$\dfrac{dy}{dx} = 0$ when $2x - 4 = 0$

So $x = 2$.

Step 3: Substitute the x-value into equation of curve to find the y-coordinate.

When $x = 2$, $y = (2 - 2)^2 + 3 = 3$

So there is a stationary point at $(2, 3)$.

Note: This confirms what you would know from work on quadratic functions and completing the square. For the curve $y = (x - p)^2 + q$, the minimum value is q and it occurs when $x = p$. This indicates that there is a minimum turning point at (p, q).

Recall:
Completing the square (Section 1.5) and Sketching curves (Section 1.13).

Second derivative

The second derivative, $\dfrac{d^2y}{dx^2}$, is obtained by differentiating $\dfrac{dy}{dx}$ with respect to x.

In function notation, if $y = f(x)$, the second derivative is written $f''(x)$.

Note:
$\dfrac{d^2y}{dx^2}$ is read as 'dee two y by dee x squared' and $f''(x)$ is read 'f double dashed x'.

Example 3.9 Given that $y = 4x^3 + 5x^2 - 3x + 1$, find   **a** $\dfrac{dy}{dx}$ **b** $\dfrac{d^2y}{dx^2}$.

Step 1: Differentiate with respect to x.

$$y = 4x^3 + 5x^2 - 3x + 1$$

a $\dfrac{dy}{dx} = 12x^2 + 10x - 3$

Step 2: Differentiate again.   **b** $\dfrac{d^2y}{dx^2} = 24x + 10$

Example 3.10 If $f(x) = x^3$, find $f''(2)$.

Step 1: Differentiate twice.

$$f(x) = x^3$$
$$f'(x) = 3x^2$$
$$f''(x) = 6x$$

Step 2: Substitute $x = 2$. $\Rightarrow$ $f''(2) = 6 \times 2 = 12$

Determining the nature of a stationary point

Here are two methods for determining the nature of a stationary point:

Method 1

Investigate the value of $\dfrac{dy}{dx}$ for x-values immediately to the left and to the right of the point.

If, as x increases,

- $\dfrac{dy}{dx}$ goes from positive to zero to negative, there is a maximum point

- $\dfrac{dy}{dx}$ goes from negative to zero to positive, there is a minimum point.

Method 2

Consider the sign of $\dfrac{d^2y}{dx^2}$ at the stationary point:

- $\dfrac{dy}{dx} = 0$ and $\dfrac{d^2y}{dx^2} > 0 \Rightarrow$ minimum turning point

- $\dfrac{dy}{dx} = 0$ and $\dfrac{d^2y}{dx^2} < 0 \Rightarrow$ maximum turning point.

Note that if $\dfrac{dy}{dx} = 0$ and $\dfrac{d^2y}{dx^2} = 0$, Method 2 fails and Method 1 is advised.

Example 3.11   **a**   Find the coordinates of the stationary points on the curve $y = x^3 - 3x + 2$.

     b   Investigate their nature.

Note:
Another method involves investigating the y-value immediately to the left and right of the stationary point.

Note:
Direction of slope near stationary point:

Maximum:

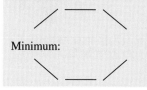

Minimum:

Tip:
Method 2 is often a quicker approach.

Tip:
Notice that the question suggests there is more than one.

Step 1: Find $\dfrac{dy}{dx}$.   **a** $y = x^3 - 3x + 2 \Rightarrow \dfrac{dy}{dx} = 3x^2 - 3$

Step 2: Put $\dfrac{dy}{dx} = 0$ and solve for x.

$$\dfrac{dy}{dx} = 0 \text{ when } 3x^2 - 3 = 0$$

$$3(x-1)(x+1) = 0$$

$$\Rightarrow \quad x = 1 \text{ or } x = -1$$

Step 3: Substitute x-value into equation of the curve.

When $x = 1$, $y = 1^3 - 3(1) + 2 = 0$

When $x = -1$, $y = (-1)^3 - 3(-1) + 2 = 4$

Tip:
Take care with negatives.

Step 4: State the coordinates.

There are stationary points at $(1, 0)$ and $(-1, 4)$.

Step 5: Find $\dfrac{d^2y}{dx^2}$.

b   Using second derivative:

$$\dfrac{dy}{dx} = 3x^2 - 3 \Rightarrow \dfrac{d^2y}{dx^2} = 6x$$

Step 6: Substitute x-value to find nature.

When $x = 1$, $\dfrac{d^2y}{dx^2} = 6(1) = 6 > 0 \Rightarrow$ minimum point.

When $x = -1$, $\dfrac{d^2y}{dx^2} = 6(-1) = -6 < 0 \Rightarrow$ maximum point.

There is a minimum turning point at $(1, 0)$ and a maximum turning point at $(-1, 4)$.

Tip:
Summarise the information, so that it is easier for the examiner to see.

Example 3.12   Find the coordinates of the stationary point of the curve $y = x^4 + 2$ and determine its nature.

Step 1: Find $\dfrac{dy}{dx}$. $y = x^4 + 2 \Rightarrow \dfrac{dy}{dx} = 4x^3$

Step 2: Put $\dfrac{dy}{dx} = 0$ and solve for x.

$\dfrac{dy}{dx} = 0$ when $4x^3 = 0$, so $x = 0$

Step 3: Substitute x-value into equation of curve.

When $x = 0$, $y = 0^4 + 2 = 2$

There is a stationary point at $(0, 2)$.

Step 4: Find second derivative.

Using the second derivative:

$$\dfrac{d^2y}{dx^2} = 12x^2$$

When $x = 0$, $\dfrac{d^2y}{dx^2} = 12(0) = 0$

The second derivative method has broken down, so use Method 1 near $x = 0$.

Step 5: Calculate the value of $\dfrac{dy}{dx}$ close to the stationary point.

When $x = -1$, $\dfrac{dy}{dx} = 4x^3 = 4(-1)^3 = -4$

When $x = 1$, $\dfrac{dy}{dx} = 4x^3 = 4(1)^3 = 4$

Tip:
You must evaluate the gradient to show that you have done the calculation.

Step 6: Draw a line representing the slope of the gradient.

	$x = -1$	$x = 0$	$x = 1$
Sign of $\dfrac{dy}{dx}$	$-$	0	$+$
Direction of gradient	\	—	/

Tip:
Do not use x-values too far away from the stationary point as you may pass another stationary point!

Step 7: State the nature of the stationary point.

There is a minimum turning point at $(0, 2)$.

3.7 Application problems

Application to determining maxima and minima.

Differentiation can be used to solve practical problems involving maximum and minimum values. In these, you consider two variables representing some measures that are related, such as the length of a rectangle and its area. Allow one to vary and use differentiation to find the maximum or minimum value of the other.

Note:
These are sometimes called optimisation problems.

Example 3.13 A carpenter is building a rectangular tabletop. In order to ensure the correct sizing he wishes to have a perimeter of 200 cm. Given that he can vary the length, he wants to know the maximum area the tabletop can have.

a By letting x cm be the length of one side of the tabletop, form an expression in terms of x, for A, the area in cm², of the tabletop.

b Use differentiation to find the maximum area of the top, verifying that you have found the maximum.

c Describe the rectangular tabletop that has maximum area.

Step 1: If applicable, draw a clear diagram, labelling all the measures and identify any unknown measures.

a

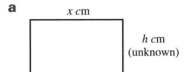

x cm

h cm (unknown)

Tip:
Let the unknown measure be h.

Step 2: Express any unknown measures in terms of x.

Perimeter $= 200$ cm

$\Rightarrow 2x + 2h = 200$

$2h = 200 - 2x$

$h = 100 - x$ ①

Tip:
Use the information given about the object to express h in terms of x.

Step 3: Form an expression for A in terms of x.

b Let the area be A cm².

$A = x \times h$

Substituting for h from ①

$A = x(100 - x)$

$= 100x - x^2$

Note:
This expression tells you how A changes for different values of x.

Step 4: Set $\dfrac{dA}{dx} = 0$ and solve for x.

$\dfrac{dA}{dx} = 100 - 2x$

$\dfrac{dA}{dx} = 0$ when $100 - 2x = 0 \Rightarrow x = 50$

Note:
This is the value of the length that gives a stationary point for A.

Step 5: Find $\dfrac{d^2A}{dx^2}$ to determine the nature of the stationary point.

$\dfrac{d^2A}{dx^2} = -2$

When $x = 50$, $\dfrac{d^2A}{dx^2} = -2 < 0 \Rightarrow$ maximum value of A.

Note:
In this case, $\dfrac{d^2A}{dx^2} < 0$ for all values of x.

Step 6: Find the corresponding A-value.

$A = 100 \times 50 - 50^2 = 2500$

The maximum area of the tabletop is 2500 cm² and this occurs when the length is 50 cm.

Step 7: Consider the dimensions of the tabletop.

c Length $= x$ cm $= 50$ cm

Width $= (100 - x)$ cm $= 50$ cm

The rectangular tabletop that gives the maximum area is in the shape of a square.

1 Find the coordinates of the stationary points of the given curves and determine their nature.

   a $y = x^2 + x^3$ **b** $y = x^3 - 3x$ **c** $y = 2x^3 + 3x^2 - 12x + 6$

2 Find the x-coordinates of the stationary points on the curve $y = (1 - x^2)(1 - 4x)$ and determine their nature.

3 A builder wishes to make a rectangular enclosure around a garden. The house is to form one of the sides; this side has length $4x$ metres. The other three sides are to be fenced with a total of 2000 m of fencing.

   Show that the area of the garden, A, is given by $A = 4000x - 8x^2$.

   Hence, find the maximum area of the garden, verifying that the value you have found is a maximum.

4 The sum of two variable positive numbers is 200.

   Let x be one of the numbers, and let the product of these two numbers be y.

   Find the maximum value of y.

5 The diagram shows a square piece of card, with sides of length 6 cm. A smaller square, of side x cm is cut from each corner as shown. The card is folded along the dotted lines to make an open box.

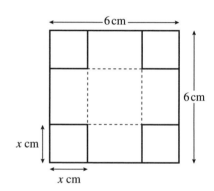

   a The volume of the box is V cm³. Show that $V = 36x - 24x^2 + 4x^3$.

   b Given that x can vary, find the values of x for which $\dfrac{dV}{dx} = 0$.

   c Show that one of the values of x gives a maximum value for V and find the maximum value of V.

6 The diagram shows a triangular prism whose cross section is a right-angled triangle with sides $3x$ cm, $4x$ cm and $5x$ cm. The length of the prism is l cm. The sum of the edges of the prism is 90 cm and it has a volume of V cm³.

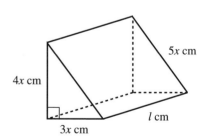

   a Show that $l = 30 - 8x$.

   b Show that $V = 12x^2(15 - 4x)$.

   c Find $\dfrac{dV}{dx}$ and hence calculate the value of x that gives a maximum volume.

   d The surface area of the prism is A cm². Express A in terms of x.

   e Show that the surface area is greatest when $x = 2\frac{1}{7}$ cm.

SKILLS CHECK **3C EXTRA** is on the CD

1 It is given that $f(x) = 2x^3 + 3x^2 + 1$.

   a Find the derivative $f'(x)$, factorising your answer.

   b Hence find the coordinates of the stationary points on the curve and determine their nature.

   c Find the values of x for which $f(x)$ is an increasing function.

2 The function f is defined for all real values of x by $f(x) = (x^2 + 4)(2x - 1)$.

   a Prove that the curve with equation $y = f(x)$ crosses the x-axis at only one point and state the x-coordinate of this point.

   b i Differentiate $f(x)$ with respect to x to obtain $f'(x)$.

   ii Hence show that the gradient of the curve $y = f(x)$ is 12 at the point where $x = 1$.

   iii Prove that the curve $y = f(x)$ has no stationary point.   [AQA (B) June 2001]

 3 An office worker can leave home at any time between 6.00 am and 10.00 am each morning. When he leaves home x **hours** after 6.00 am ($0 \leqslant x \leqslant 4$), his journey time to the office is y **minutes**, where
$y = x^4 - 8x^3 + 16x^2 + 8$.

   a Find $\dfrac{dy}{dx}$.   *(3 marks)*

   b Find the **three** values of x for which $\dfrac{dy}{dx} = 0$.   *(4 marks)*

   c Show that y has a maximum value when $x = 2$.   *(3 marks)*

   d Find the time at which the office worker arrives at the office on a day when his journey time is a maximum.   [AQA (A) Nov 2002]

4 Small trays are to be made from rectangular pieces of card. Each piece of card is 8 cm by 5 cm and the tray is formed by removing squares of side x cm from each corner and folding the remaining card along the dashed lines, as shown in the diagram, to form an open-topped box.

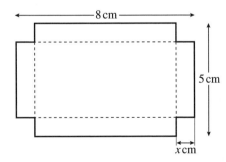

   a Explain why $0 < x < 2.5$.

   b Show that the volume, V cm^3, of a tray is given by
   $V = 4x^3 - 26x^2 + 40x$.

   c Find the value of x for which $\dfrac{dV}{dx} = 0$.

   d Calculate the greatest possible volume of a tray.   [AQA (A) Jan 2001]

5 An open-topped box has height h cm and a square base of side x cm.

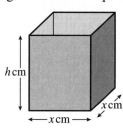

The box has capacity V cm³. The area of its **external** surface, consisting of its horizontal base and four vertical faces, is A cm².

a Find expressions for V and A in terms of x and h.

b It is given that $A = 3000$.

   i Show that $V = 750x - \frac{1}{4}x^3$.

   ii Find the positive value of x for which $\dfrac{\mathrm{d}V}{\mathrm{d}x} = 0$, giving your answer in surd form.

   iii Hence find the maximum possible value of V, giving your answer in the form $p\sqrt{10}$, where p is an integer.
     [You do not need to show that your answer is a maximum.]   [AQA (A) June 2001]

6 The sketch shows the curve $y = x^2 - 3x + 1$.
The point P has x coordinate 1 and lies on the curve.

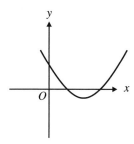

a Find $\dfrac{\mathrm{d}y}{\mathrm{d}x}$ when $x = 1$.

b Hence find the gradient of the normal at P.

c Find the equation of the normal at P.

d The normal at P meets the curve again at Q. Find the x-coordinate of Q.

 7 The normal at the point $(0, 3)$ on the curve $y = x^2 - 2x + 3$ meets the curve again at A. Find the coordinates of A.

4 Integration

4.1 The reverse of differentiation

Indefinite integration as the reverse of differentiation.

The reverse process of differentiation is called **integration**.

This means that, given $\dfrac{dy}{dx}$, you integrate to find y; given $f'(x)$, you integrate to find $f(x)$.

Indefinite integration

You will have noticed when differentiating that more than one function can have the same derivative, for example:

$$y = 4x^2 \qquad \Rightarrow \frac{dy}{dx} = 8x$$

$$y = 4x^2 - 6 \quad \Rightarrow \frac{dy}{dx} = 8x$$

In fact, for any constant c:

$$y = 4x^2 + c \quad \Rightarrow \frac{dy}{dx} = 8x$$

Recall:
Differentiation of polynomials.
If $y = ax^n$, $\dfrac{dy}{dx} = nax^{n-1}$
(Section 3.2).

When integrating, you take this into account by including a constant, called an **integration constant**. All possible values of the constant give a **family** of solutions, known as the **general solution**.

Recall:
The derivative of a constant is zero (Section 3.2).

This process is called **indefinite integration**. There are several ways of writing it, including:

$$\frac{dy}{dx} = 8x \Rightarrow y = 4x^2 + c$$

$$f'(x) = 8x \Rightarrow f(x) = 4x^2 + c$$

$$\int 8x\,dx = 4x^2 + c$$

Note:
Often c is used for the integration constant. It can take any value, positive, negative or zero.

Note:
$\int 8x\,dx$ is read 'the integral of $8x$, with respect to x'.

4.2 Polynomial integration

Integration of polynomials.

In Core 1 you have to be able to integrate constant functions such as $f(x) = 3$, and functions containing terms in x^n, where n is a positive integer, such as $f(x) = 3x^5$.

Note:
Other values of n are considered in Core 2.

To integrate x^n, reverse the differentiation process as follows:

- increase the power by one

- divide by the new power.

For positive integer values of n:

$$\int x^n\,dx = \frac{1}{n+1}x^{n+1} + c$$

Note:
You can write
$$\int x^n\,dx = \frac{x^{n+1}}{n+1} + c.$$

For constant function k: $\int k\,dx = kx + c$

If a is a constant:

$$\int a\mathrm{f}(x)\,dx = a\int \mathrm{f}(x) \Rightarrow \int ax^n\,dx = \frac{a}{n+1}x^{n+1} + c$$

Example 4.1 Find   **a** $\int x^7\,dx$ **b** $\int 3\,dx$ **c** $\int 3x^4\,dx$

Step 1: Apply the integration formula.

a $\int x^7\,dx = \frac{1}{8}x^8 + c$

b $\int 3\,dx = 3x + c$

c $\int 3x^4\,dx = \frac{3}{5}x^5 + c$

Example 4.2 Find $\int \frac{2}{3}x^6\,dx$.

Step 1: Apply the appropriate integration formula.

$$\int \tfrac{2}{3}x^6\,dx = \tfrac{2}{3} \times \int x^6\,dx$$
$$= \tfrac{2}{3} \times \tfrac{1}{7}x^7 + c$$
$$= \tfrac{2}{21}x^7 + c$$

To integrate a **polynomial** in x containing several terms, such as $x^3 + 5x^2 + 2x - 1$, integrate term by term, using the rule

$$\int (\mathrm{f}(x) \pm \mathrm{g}(x))\,dx = \int \mathrm{f}(x)\,dx \pm \int \mathrm{g}(x)\,dx.$$

Example 4.3 Find $\int (x^3 + 5x)\,dx$.

Step 1: Integrate term by term.

$$\int (x^3 + 5x)\,dx = \tfrac{1}{4}x^4 + \tfrac{5}{2}x^2 + c$$

Example 4.4 Given that $\frac{dy}{dx} = 5x^4 - 6x + 3$, express y in terms of x.

Step 1: Integrate term by term.

$$y = \int (5x^4 - 6x + 3)\,dx$$

Step 2: Simplify where possible.

$$= \tfrac{5}{5}x^5 - \tfrac{6}{2}x^2 + 3x + c$$
$$= x^5 - 3x^2 + 3x + c$$

Example 4.5 Given that $\mathrm{f}'(x) = x^2(3 - 2x^4)$, find $\mathrm{f}(x)$.

Step 1: Rewrite so that each term is in the form ax^n.

$$\mathrm{f}'(x) = x^2(3 - 2x^4)$$
$$= 3x^2 - 2x^6$$

Step 2: Integrate term by term.

Step 3: Simplify where possible.

$$\mathrm{f}(x) = \int (3x^2 - 2x^6)\,dx$$
$$= \tfrac{3}{3}x^3 - \tfrac{2}{7}x^7 + c$$
$$= x^3 - \tfrac{2}{7}x^7 + c$$

Sometimes you have additional information enabling you to find a specific value for the integration constant. In this case you can give a **particular solution**, rather than a general one.

Example 4.6 A curve with gradient function $2x - 6x^2$ goes through the point $P(2, -10)$. Find the equation of the curve.

Recall:
Gradient function (Section 3.1).

Step 1: Write $\dfrac{dy}{dx}$ for the gradient function and integrate.

$$\frac{dy}{dx} = 2x - 6x^2$$

$$y = x^2 - 2x^3 + c$$

Note:
The general solution contains the integration constant c.

Step 2: Substitute the coordinates of the given point to find c.

At P, $x = 2$ and $y = -10$.

Substituting into $y = x^2 - 2x^3 + c$ gives

$$-10 = 2^2 - 2 \times 2^3 + c$$

$$-10 = 4 - 16 + c$$

$$c = 2$$

The equation of the curve is $y = x^2 - 2x^3 + 2$.

Note:
Use the fact that P lies on the curve to find the value of c.

Note:
This is the particular solution.

SKILLS CHECK 4A: Indefinite integration of polynomials

1 Find

   a $\displaystyle\int 4x^2\,dx$ **b** $\displaystyle\int \tfrac{2}{5}x^2\,dx$ **c** $\displaystyle\int -6\,dx$.

2 Find y in terms of x if:

   a $\dfrac{dy}{dx} = x^5 + 3x^2$ **b** $\dfrac{dy}{dx} = 2x^4 - \tfrac{1}{2}x^3$ **c** $\dfrac{dy}{dx} = x(2x + 7)$

   d $\dfrac{dy}{dx} = 4 - 3x$ **e** $\dfrac{dy}{dx} = -\tfrac{2}{5}x^3 - 1$ **f** $\dfrac{dy}{dx} = 2c$

 3 a Expand $(2x + 3)(x - 4)$.   **b** Find $\displaystyle\int (2x + 3)(x - 4)\,dx$.

4 a Given that $\dfrac{dA}{dt} = 4t^7$, find A in terms of t.

   b Given that $A = 0$ when $t = 1$, find the value of A when $t = 2$.

5 A curve passes through the origin and has gradient function $5x - 3$. Find the equation of the curve.

6 A curve passes through $P(1, 0)$ and has gradient function $3x^2 + 2x + 4$. Find the equation of the curve.

7 Find an expression for $f(x)$ if $f'(x) = 4 - x^2$ and $f(1) = 0$.

8 Find   **a** $\displaystyle\int (2x + 3)^2\,dx$ **b** $\displaystyle\int (2x + 3)(2x - 3)\,dx$.

SKILLS CHECK 4A EXTRA is on the CD

4.3 Definite integrals

Evaluation of definite integrals.

A **definite integral** has the form $\int_a^b f(x)dx$, where a and b are the **limits** of integration.

Note:
a is the lower limit and b is the upper limit.

To **evaluate** a definite integral:

* integrate $f(x)$ but omit the integration constant, c,

* substitute the upper limit,

* subtract from this the value obtained when the lower limit is substituted.

Note:
You have to work out a specific value rather than give it as a function of x. It does not contain an integration constant c.

Example 4.7 Evaluate $\int_3^4 2x\,dx$.

Step 1: Integrate the function, simplifying if possible.

Step 2: Substitute the limits, upper limit first.

Step 3: Calculate the numerical value.

$$\int_3^4 2x\,dx = \left[\tfrac{2}{2}x^2\right]_3^4$$

$$= \left[x^2\right]_3^4 = 4^2 - 3^2$$

$$= 16 - 9 = 7$$

Tip:
When you have integrated, use square brackets and put the limits on the right.

Note:
The answer is a number, not a function of x.

Example 4.8 Evaluate $\int_{-3}^{-2}(5 - 4x)dx$.

Step 1: Integrate the function, simplifying if possible.

Step 2: Substitute the limits, upper limit first.

Step 3: Calculate the numerical value.

$$\int_{-3}^{-2}(5 - 4x)dx = \left[5x - \tfrac{4}{2}x^2\right]_{-3}^{-2}$$

$$= \left[5x - 2x^2\right]_{-3}^{-2}$$

$$= 5(-2) - 2(-2)^2 - (5(-3) - 2(-3)^2)$$

$$= -10 - 8 - (-15 - 18)$$

$$= -18 - (-33) = 15$$

Note:
Take great care with the negative numbers.

To simplify the working, it is often a good idea to take out any numerical factors before substituting the limits.

Note:
You are not allowed to use a calculator in Core 1.

Example 4.9 Evaluate $\int_2^4 5x^2\,dx$.

Step 1: Integrate the function.

Step 2: Take out any numerical factors.

Step 3: Substitute the limits, upper limit first.

Step 4: Calculate the values.

Step 5: Tidy any fractions.

$$\int_2^4 5x^2\,dx = \left[\tfrac{5}{3}x^3\right]_2^4$$

$$= \tfrac{5}{3}\left[x^3\right]_2^4$$

$$= \tfrac{5}{3}(4^3 - 2^3)$$

$$= \tfrac{5}{3}(64 - 8) = \tfrac{5}{3} \times 56$$

$$= \tfrac{280}{3} = 93\tfrac{1}{3}$$

Note:
There is a factor of $\frac{5}{3}$.

Interpretation of the definite integral as the area under a curve.

Area under a curve

Integration can be used to find areas bounded by lines and curves. This is often referred to as finding the area 'under' a curve.

Consider the area of a region bounded by a curve, the x-axis and the lines $x = a$ and $x = b$.

If the area is *above* the x-axis:

$$\text{Area} = \int_a^b f(x)dx$$

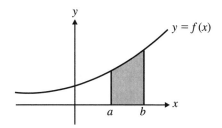

> **Note:**
> This can be written $\int_a^b y dx$.

If the area is *below* the x-axis, the value of $\int_a^b f(x)dx$ is negative.

The area is found by taking the positive value of the number calculated.

$$\text{Area} = \left| \int_a^b f(x)dx \right|$$

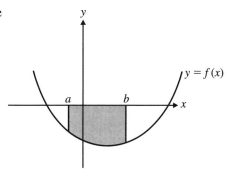

> **Note:**
> The straight lines mean that you take the positive value. This is called the modulus.

Example 4.10 Find the area of the region enclosed by the curve $y = x^2 + 2$, the x-axis and the lines $x = 1$ and $x = 2$.

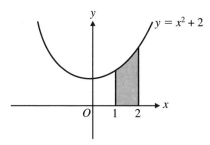

Step 1: Use the area formula with $a = 1$ and $b = 2$.

$$\int_a^b f(x)dx = \int_1^2 (x^2 + 2)dx$$

Step 2: Integrate term by term.

$$= \left[\tfrac{1}{3}x^3 + 2x \right]_1^2$$

Step 3: Substitute the limits.

$$= \tfrac{1}{3} \times 2^3 + 2 \times 2 - (\tfrac{1}{3} \times 1^3 + 2 \times 1)$$

$$= \tfrac{8}{3} + 4 - (\tfrac{1}{3} + 2)$$

Step 4: Work out the values.

$$= 2\tfrac{2}{3} + 4 - 2\tfrac{1}{3}$$

$$= 4\tfrac{1}{3}$$

The area is $4\tfrac{1}{3}$ square units.

> **Note:**
> The area is above the curve, so the integral will give a positive number.

> **Tip:**
> Be careful with the fractions. It is a good idea to write down all the working.

Example 4.11 The diagram shows a sketch of the curve $y = x(x - 3)$.

Find the area enclosed between the curve and the x-axis.

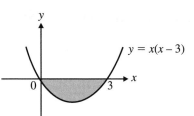

Tip:
Notice that the area is below the x-axis.

Step 1: Use the formula with $a = 0$ and $b = 3$.

$$\int_a^b f(x)dx = \int_0^3 x(x - 3)dx$$

Tip:
Write the expression in index form before you integrate.

Step 2: Expand the brackets and integrate term by term.

$$= \int_0^3 (x^2 - 3x)dx$$

$$= \left[\tfrac{1}{3}x^3 - \tfrac{3}{2}x^2\right]_0^3$$

Step 3: Substitute the limits and calculate the value.

$$= \tfrac{1}{3} \times 3^3 - \tfrac{3}{2} \times 3^2 - (0 - 0)$$

$$= 9 - 13\tfrac{1}{2}$$

$$= -4\tfrac{1}{2}$$

Tip:
It is helpful to put in the zeros to show that you have substituted for the lower limit.

Note:
As expected, the value of the integral is negative as the area lies below the x-axis.

Step 4: Find the modulus. $\text{Area} = \left| \int_a^b f(x)dx \right| = \left| -4\tfrac{1}{2} \right| = 4\tfrac{1}{2}$ square units.

Area enclosed between a line and a curve

Example 4.12 The curve $y = x^2$ and the line $y = 2 - x$ intersect at $A(-2, 4)$ and $B(1, 1)$. Find the area enclosed between the line and the curve.

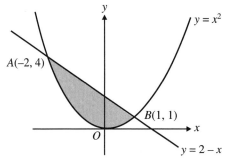

Step 1: Find the area under the line from $x = -2$ to $x = 1$.

First find the area under the line between $x = -2$ and $x = 1$.

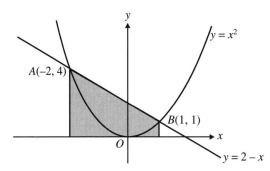

Note:
You could evaluate
$\int_{-2}^{1} (2 - x)dx$ to find the area under the line.

Area of trapezium $= \tfrac{1}{2}(4 + 1) \times 3$

$= 7\tfrac{1}{2}$ square units

Recall:
Area of a trapezium $= \tfrac{1}{2}$ (sum of parallel sides) $\times$ perpendicular distance between them.

Now find the area under the curve between $x = -2$ and $x = 1$.

Step 2: Find the area under the curve from $x = -2$ to $x = 1$.

Area under the curve:

$$\int_a^b f(x)dx = \int_{-2}^1 x^2 dx$$

$$= \left[\tfrac{1}{3}x^3\right]_{-2}^1$$

$$= \tfrac{1}{3}(1^3 - (-2)^3)$$

$$= \tfrac{1}{3} \times 9$$

$$= 3$$

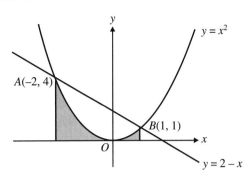

Area under curve = 3 square units

Step 3: Subtract the area under the curve from the area under the line.

Required area $= 7\tfrac{1}{2} - 3$

$\qquad\qquad = 4\tfrac{1}{2}$ square units

SKILLS CHECK **4B: Definite integration of polynomials and areas under curves**

1 Evaluate the following definite integrals:

a $\displaystyle\int_1^2 3x^2 dx$ **b** $\displaystyle\int_0^2 x^4 dx$ **c** $\displaystyle\int_2^3 \tfrac{1}{2}x dx$

 2 a Evaluate $\displaystyle\int_{-2}^2 (x + 3)dx$.   **b** Evaluate $\displaystyle\int_{-2}^{-1} (x^3 + x)dx$.

3 Writing the functions in index form first, evaluate the following:

a $\displaystyle\int_0^1 (x - 1)^2 dx$ **b** $\displaystyle\int_{-2}^2 (2 - x)(2 + x)dx$

4 The diagram shows a sketch of the curve $y = 4x^3 + 1$.

Find the area of the region enclosed by the curve, the x-axis and the lines $x = 1$ and $x = 2$.

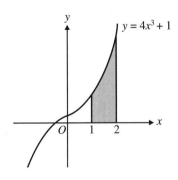

5 The curve $y = (2 - x)(1 + x)$ crosses the x-axis at A and B.

a Find the coordinates of A and B.

b Find the area enclosed by the curve and the x-axis.

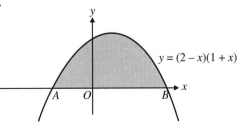

6 a Sketch the curve $y = 2x(1 - x)$.

b Find the area enclosed by the curve and the x-axis.

7 **a** Solve the simultaneous equations:
$$y = x^2 + 2$$
$$x + y = 4$$

b The curve $y = x^2 + 2$ and the line $x + y = 4$ intersect at P and Q.

    i Write down the coordinates of P and Q.

    ii Find the area enclosed between the line and the curve.

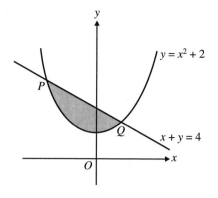

SKILLS CHECK **4B EXTRA** is on the CD

Examination practice Integration

 1 The gradient of a curve at the point (x, y) is given by $\dfrac{dy}{dx} = 3x^2 - x$. Use integration to find the equation of the curve, given that the curve passes through the point $(2, 1)$.

2 It is given that $y = 4x^3 + 1$.

  a Find $\dfrac{dy}{dx}$.

  b **i** Find $\displaystyle\int y\,dx$

      ii Hence evaluate $\displaystyle\int_1^2 y\,dx$.

3 Show that $\displaystyle\int_1^2 (\tfrac{3}{4}x^2 - \tfrac{1}{2}x)\,dx = 1$.

4 The diagram shows the graph of $y = x^3 - x$, $x \geqslant 0$.
The points on the graph for which $x = 1$ and $x = 2$ are labelled A and B, respectively.

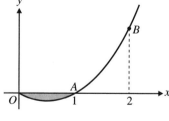

  a Find the y-coordinate of B and hence find the equation of the straight line AB, giving your answer in the form $ax + by + c = 0$.

  b Find, by integration, the area of the shaded region.

[AQA (A) Nov 2002]

5 The curve with equation $y = 12 + 4x - x^2$ cuts the y-axis at A, the positive x-axis at B and the negative x-axis at C as shown in the diagram. The point O is the origin and the maximum point of the curve is M. The shaded region R is bounded by the line AB and the curve.

The point B has coordinates $(6, 0)$.

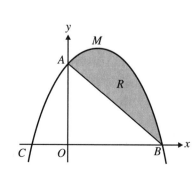

  a Show that $x = 2$ at the point M.

  b Find the coordinates of C.

  c Show that triangle OAB and the region R have equal areas.

[AQA (B) June 2001]

6 The diagram shows a part of the graph of $y = x - 2x^4$.

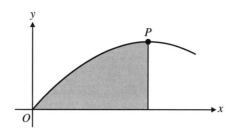

a i Find $\dfrac{dy}{dx}$.

   ii Show that the x-coordinate of the stationary point P is $\frac{1}{2}$.

   iii Find the y-coordinate of P.

b i Find $\displaystyle\int (x - 2x^4)dx$.

   ii Hence find the area of the shaded region.   [AQA (A) Jan 2003]

7 The line $y = 2x + 5$ intersects the curve $y = x^2 + 2x + 2$ at the points P and Q, as shown in the diagram.

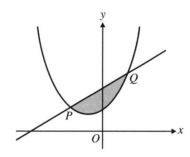

a Find the coordinates of P and Q, giving your answers in surd form.

b Find the area of the shaded region, giving your answer in surd form.   [AQA (A) Jan 2001]

8 The diagram shows the graph of $y = 12 - 3x^2$ and the tangent to the curve at the point $P(2, 0)$. The region enclosed by the tangent, the curve and the y-axis is shaded.

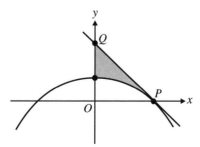

a Find $\displaystyle\int_0^2 (12 - 3x^2)dx$.

b i Find the gradient of the curve $y = 12 - 3x^2$ at the point P.

   ii Find the coordinates of the point Q where the tangent at P crosses the y-axis.

c Find the area of the shaded region.   [AQA (A) May 2002]

9 The function f is defined for all values of x by $f(x) = x^3 - 7x^2 + 14x - 8$.

It is given that $f(1) = 0$ and $f(2) = 0$.

a Find the values of $f(3)$ and $f(4)$.

b Write $f(x)$ as a product of **three** linear factors.

c The diagram shows the graph of $y = x^3 - 7x^2 + 14x - 8$.

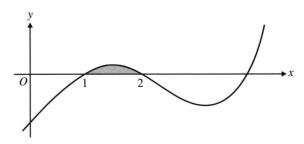

i Find $\dfrac{dy}{dx}$.

ii State, giving a reason, whether the function f is increasing or decreasing at the point where $x = 3$.

iii Find $\displaystyle\int (x^3 - 7x^2 + 14x - 8)dx$.

iv Hence find the area of the shaded region enclosed by the graph of $y = f(x)$, for $1 \leqslant x \leqslant 2$, and the x-axis. [AQA (A) May 2003]

Practice exam paper

1 Express each of the following in the form $a + b\sqrt{2}$, where a and b are integers.

   a $\sqrt{8}(7 - \sqrt{2})$ *(3 marks)*

   b $\dfrac{21}{3 + \sqrt{2}}$ *(3 marks)*

2 The points $A(1, 3)$ and $B(3, -1)$ lie on the line with equation $y = -2x + 5$.
The mid-point of AB is M.

   a   Find the coordinates of M.   *(2 marks)*

   b   **i**   Find an equation of the line which is perpendicular to AB and passes through
               the point M.   *(3 marks)*

          ii   Deduce that the perpendicular bisector of AB passes through the origin $(0, 0)$.   *(1 mark)*

3 **a**   Solve the inequality
$$4(x - 2) \geqslant 7 - 2(x - 3).$$
                                                                        (3 marks)

   b   Given that the equation
$$kx^2 + kx + 2 = 0,$$
       where k is a constant, has no real roots, find the values of k.   *(4 marks)*

4 The diagram shows the graph of $y = x^4 - 2x^2 + 1$.

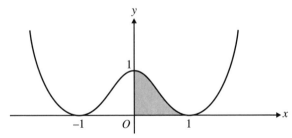

   a   Show that the gradient of the curve at the point $(2, 9)$ is 24.   *(3 marks)*

   b   **i**   Find $\int (x^4 - 2x^2 + 1)\,dx$.   *(3 marks)*

          ii   Hence find the area of the shaded region bounded by the curve and the positive
               coordinate axes.   *(3 marks)*

5 A polynomial is given by
$$f(x) = x^3 + kx^2 - 6x - 8.$$

   a   Given that when $f(x)$ is divided by $(x - 1)$ the remainder is -10, show that $k = 3$.   *(2 marks)*

   b   Show that $(x + 1)$ is a factor of $f(x)$.   *(2 marks)*

   c   Find the value of $f(2)$ and hence, or otherwise, write $f(x)$ as a product of three
       linear factors.   *(4 marks)*

6 A curve C has equation $y = (x + 2)^2 - 4$.

    a Describe a geometrical transformation by which C can be obtained from the curve with equation $y = x^2$. *(3 marks)*

    b **i** Solve the equation $(x + 2)^2 = 4$. *(2 marks)*

       ii Sketch the curve C. Indicate the coordinates of any points of intersection with the coordinate axes. *(2 marks)*

      iii Write down the coordinates of the vertex of the curve C. *(2 marks)*

    c **i** Show that the x-coordinates of the points of intersection of the line with equation $y = 2x + 3$ and the curve C satisfy the equation
$$x^2 + 2x - 3 = 0.$$ *(3 marks)*

       ii Hence calculate the coordinates of the points of intersection of the line $y = 2x + 3$ and the curve C. *(3 marks)*

7 A curve has equation
$$y = 72 + 36x - 3x^2 - 4x^3.$$
The curve has two stationary points.

    a Find $\dfrac{dy}{dx}$. *(3 marks)*

    b Find the values of x for which $\dfrac{dy}{dx} = 0$. *(3 marks)*

    c **i** Find $\dfrac{d^2y}{dx^2}$. *(2 marks)*

       ii Find the value of $\dfrac{d^2y}{dx^2}$ at each of the stationary points. *(2 marks)*

    d Write down the coordinates of the minimum point of the curve. *(2 marks)*

8 A circle has equation
$$x^2 + y^2 - 2x + 4y = 95.$$

    a By completing the square, express this equation in the form
$$(x - a)^2 + (y - b)^2 = d.$$ *(3 marks)*

    b The point C is the centre of the circle. Write down the coordinates of C and show that the radius of the circle is 10. *(2 marks)*

    c The points $P(7, 6)$ and $Q(-7, 4)$ lie on the circle.

       i Find the equation of the normal to the circle at P. *(3 marks)*

       ii Show that the equation of the tangent to the circle at Q is $3y = 4x + 40$ and deduce that the normal to the circle at P and the tangent to the circle at Q never intersect. *(4 marks)*

Answers

1 a $5\sqrt{2}$ **b** $4\sqrt{2}$ **c** $7\sqrt{2}$ **d** $7\sqrt{3}$ **e** $7\sqrt{7} - 4\sqrt{5}$
2 a $11\sqrt{3}$ **b** $2\sqrt{2}$ **c** $5 - 2\sqrt{6}$ **d** 50 **e** $2\sqrt{2}$
 f 3 **g** $\dfrac{\sqrt{3}}{3}$ **h** $\sqrt{6} - 2$
3 a $12 + 2\sqrt{11}$ **b** 10 **c** $\dfrac{6}{5} + \dfrac{\sqrt{11}}{5}$
4 $p = \dfrac{3}{2}$
5 $4\sqrt{3}, 2\sqrt{3}$ **b** $6\sqrt{3}$
6 a $8 + 2\sqrt{7}$ **b** $\dfrac{4}{3} + \dfrac{1}{3}\sqrt{7}$
7 a $p = 7, q = 1$ **b** $r = \dfrac{7}{11}, s = \dfrac{1}{11}$

1 a $x(x + 5)$ **b** $(x - 1)(x - 1)$ **c** $(a + 4)(a - 4)$
 d $(x + 1)(x - 6)$ **e** $(x + 15)(x - 2)$ **f** $2x(x - 4)$
2 a $(2x + 3)(x + 2)$ **b** $(5x + 1)(x - 3)$ **c** $x(x + 3)(x - 7)$
 d $(3y - 2)(y + 2)$ **e** $4(1 - 2x)(5x + 3)$ **f** $(2x + 5)(2x - 5)$
3 a $(x + 3)^2 - 1$ **b** $(x - 6)^2 - 39$ **c** $\left(x + \dfrac{5}{2}\right)^2 - \dfrac{33}{4}$
4 a $2\left(x + \dfrac{7}{4}\right)^2 - \dfrac{1}{8}$ **b** $5\left(x - \dfrac{7}{5}\right)^2 - \dfrac{64}{5}$ **d** $3\left(y + \dfrac{2}{3}\right)^2 - \dfrac{16}{3}$
 e $-40\left(x + \dfrac{1}{20}\right)^2 + \dfrac{121}{10}$ **f** $4x^2 - 25$
5 a $p = 2, q = 3$ **b** $(2, 3)$ **c** Minimum **d** $x = 2$
6 a $4(x + 1)^2 - 3$ **b** $(-1, -3)$, minimum **c** $x = -1$
7 a $18 - (x + 2)^2$ **b** 18 **c** $(-2, 18)$

1 a $x = 2, x = -3$ **b** $x = \frac{1}{4}, x = -\frac{3}{2}$ **c** $x = 0, x = -5$
2 a $x = -5, x = -1$ **b** $x = 8, x = 3$ **c** $x = 0, x = 6$
 d $x = -1, x = 6$ **e** $x = -2, x = 3$ **f** $x = \pm 6$
3 a $\left(x + \dfrac{3}{2}\right)^2 - \dfrac{29}{4}$ **b** $x = \dfrac{-3 \pm \sqrt{29}}{2}$
4 a $2\left(x - \dfrac{3}{4}\right)^2 - \dfrac{25}{8}$ **b** $x = -\dfrac{1}{2}, x = 2$
5 $x = \dfrac{-3 \pm \sqrt{69}}{10}$
6 a $x = -1, x = 4$ **b** $x = \dfrac{7 \pm \sqrt{13}}{4}$
7 $x = -\dfrac{1}{3} \pm \dfrac{1}{3}\sqrt{13}$
8 a 2 **b** 2 **c** 1 (repeated) **d** 2
9 -56

1 a $x = 4, y = 1$ **b** $x = 3, y = 2$ **c** $a = 3, b = -2$
2 a $x = -6, y = 4$ **b** $a = 11, b = 3$ **c** $p = -9, q = 3$
3 b $x = 4, y = -\frac{2}{3}$
4 a $x = 2, y = 8$ or $x = -6, y = 16$
 b $x = \frac{2}{3}, y = 2$ or $x = -1, y = -3$
 c $x = 0, y = 1$ or $x = -1, y = 0$
5 width $= 3$ cm, length $= 6$ cm
6 b $2a + b = 4$ **c** $a = 3, b = -2$

1 a $x > 3$ **b** $x > -1$ **c** $x \geqslant 8\frac{2}{3}$ **d** $y > 1$
 e $x > 6$ **f** $x \leqslant 14.5$
2 a $3 < x < 9$ **b** $-6 \leqslant x \leqslant 10$
3 $a = 3$
4 a $y < -2, y > 2$ **b** $-7 \leqslant x \leqslant 7$ **c** $x \leqslant -\sqrt{5}, x \geqslant \sqrt{5}$
 d $-3 < x < 3$ **e** $x < -1, x > 3$
 f $-2 - \sqrt{5} \leqslant x \leqslant -2 + \sqrt{5}$
5 a $p = 2, q = -9$ **b** $x \leqslant -5, x \geqslant 1$
6 a $-4 < x < 3$ **b** $x \leqslant -2\frac{1}{2}, x \geqslant \frac{2}{3}$ **c** $x \leqslant -5, x \geqslant 4$
 d $-5 < p < -2$ **e** $x \leqslant -2, x \geqslant \frac{3}{2}$
7 a $p = 2, q = 10$ **c** $2 - \sqrt{10} < x < 2 + \sqrt{10}$
8 a $-6 < k < 6$ **b** $-4 \leqslant k \leqslant 4$
9 a $k^2 - 16$ **b** $k < -4, k > 4$

1 a $-x^2 - 18x + 5$ **b** $4x^3 + 14x^2 - 7x + 4$
 c $x^3 - 5x^2 - 9x + 45$ **d** $2x^3 + 3x^2 - 23x - 12$
2 a $a = 2, b = 5$ **b** $a = 3, b = -1, c = 5$
 c $a = -4, b = -\frac{3}{2}, c = -2$ or $a = 3, b = 2, c = -2$

1 a $\frac{1}{2}, 2, -\frac{1}{3}$ **b** $\pm 2, \pm 3$ **c** $\frac{1}{4}, -\frac{4}{3}, 2$ **d** $-3, 1$
2 a $1, 2, -3$ **b** $3, -2 \pm \sqrt{2}$ **c** $-1, -2, -3$
3 a -1 **b** 8 **c** $0, (x - 1)$ is a factor
4 a $x^2 - 3x - 5 - \dfrac{18}{x - 3}$ **b** $x^2 + 5x - 6 + \dfrac{7}{x + 1}$
 c $2x^2 + 5x + 6 + \dfrac{17}{x - 2}$
5 $2, 2 \pm \sqrt{11}$
6 $(x + 1)^3$
7 $a = 2, b = 4$
8 $a = 4, b = -1$

2 a $(1 - 2x)(x + 4)$
4 a $(x + 6)(x - 4), x = -6, x = 4$ **b** $(x + 1)^2 - 25$ **d** $x = -1$
6 b $(x - 2)(2x + 1)(x + 1)$ **c** $(0, -2)$
 d $(-1, 0), (-0.5, 0), (2, 0)$

1 a i $\binom{0}{3}$ **ii** $(0, 3), 3$ **b i** $\binom{0}{-2}$ **ii** $(0, -2), -2$
 c i $\binom{0}{1}$ **ii** $(0, 1), 1$ **d i** $\binom{-2}{0}$ **ii** $(-2, 0), 4$
 e i $\binom{1}{0}$ **ii** $(1, 0), 1$ **f i** $\binom{-4}{0}$ **ii** $(-4, 0), 16$
 g i $\binom{-3}{1}$ **ii** $(-3, 1), 10$ **h i** $\binom{2}{-1}$ **ii** $(2, -1), 3$
 i i $\binom{-1}{-2}$ **ii** $(-1, -2), -1$
2 a i $\binom{-2}{0}$ **ii** $(-2, 0), 2$ **b i** $\binom{0}{2}$ **ii** $(0, 2), 5$
 c i $\binom{-1}{3}$ **ii** $(-1, 3), 3$ **d i** $\binom{2}{-1}$ **ii** $(2, -1), 2$
3 a $(x - 1)^2 + (y - 2)^2 = 4$ **b** $(x + 1)^2 + (y + 1)^2 = 9$
4 a $y = (x - 3)^2 + 1$ **b** $\binom{3}{1}, (3, 1)$
 c Graph doesn't cross x-axis

Exam practice 1 (page 28)

1 a $4 + \sqrt{3}$ **b** $8 + 2\sqrt{3}$

2 a $2\sqrt{2} + 3$ **b** $x < \dfrac{2 + 2\sqrt{2}}{\sqrt{2} - 1}$

3 a $y < -3$ **b** $x > 2, x < -6$

4 a i $3\sqrt{5}$ **ii** $4\sqrt{5}$ **b** $7\sqrt{5}$

5 $x = 3, y = 0; x = \frac{1}{2}, y = 5$

6 $x = 3, y = -1; x = 1, y = 1$

7 $x = -1, y = 1; x = \frac{3}{2}, y = -\frac{1}{4}$

8 a $(x + 2)^2 - 9$ **b** $x < -5, x > 1$

9 a i $x = -2 \pm \frac{1}{2}\sqrt{2}$ **ii** $x < -2 - \frac{1}{2}\sqrt{2}, \ x > -2 + \frac{1}{2}\sqrt{2}$

     b $2(x + 2)^2 - 1$ **c i** -1 **ii** -2

10 a i $a = 6, b = -25$ **ii** -25

     b $k = 4$ or $\frac{4}{9}$

11 a $x = -8 \pm \frac{3}{2}\sqrt{2}$ **b i** $m = 8, n = -9$ **ii** -9

12 b $k = -6$ or 18 **c** Line is tangent to curve

13 a $a = -3, b = -2$ **b** $3 - \sqrt{2} < x < 3 + \sqrt{2}$

14 a $0, -6$ **b** $(x - 1)(x + 1)(x - 4)$

15 a 0 **b** $(x - 2)$ or $(x + 1)$ or $(x + 4)$ **c** $(x - 2)(x + 1)(x + 4)$

16 a $0, -4$ **b** $(x - 1)(x + 2)(x + 3)$

17 a $a = 6, b = 11, c = 3$ **b** $x = -1, -\frac{1}{3}, -\frac{3}{2}$

18 b $(x - 1)(2x - 3)(x + 2)$

19 $a = -3, b = 0$

20 $a = -1, b = 2$

21 a $(x - 4)^2 - 3$ **b** $\begin{pmatrix} 4 \\ -3 \end{pmatrix}$ **c** $(4, -3), (0, 13)$

22 a $(x + 1)^2 + 2$ **b** min 2 at $x = -1$ **c** $\begin{pmatrix} 1 \\ -2 \end{pmatrix}$

23 a $\begin{pmatrix} -7 \\ 7 \end{pmatrix}$ **b** centre $(-7, 7)$, radius 7

SKILLS CHECK 2A (page 34)

1 a $3x - y - 2 = 0$ **b** $3x + 2y - 6 = 0$

   c $2x + 3y - 20 = 0$ **d** $6x + 5y - 20 = 0$

2 a $y = \frac{4}{5}x - \frac{8}{5}$ **b** Gradient $= -\frac{2}{3}$, y-intercept $= 2$

3 a i -1 **ii** $(\frac{1}{2}, \frac{5}{2})$ **iii** $7\sqrt{2}$

   b i $\frac{5}{3}$ **ii** $(1, 1)$ **iii** $2\sqrt{34}$

   c i $\frac{1}{3}$ **ii** $(\frac{1}{2}, -\frac{3}{2})$ **iii** $\sqrt{10}$

4 a $y = \frac{3}{5}x + \frac{2}{5}$ **b** $y = -3x + 12$ **c** $y = -x + 3$

6 b $(-3, -1)$ **c** 15 units^2

SKILLS CHECK 2B (page 36)

1 a Perpendicular   **b** Parallel   **c** Neither

3 $y = \frac{2}{3}x + 3$

4 $y = -\frac{5}{4}x - 2$

5 $y = 5x - 16$

6 $y = 3x - 7$

7 $5x + 3y + 19 = 0$

8 $x + 5y - 12 = 0$

9 a $\dfrac{5}{2}$ **b** $y = \frac{5}{2}x - \frac{29}{2}$ **c** $-\dfrac{1}{2}$ **d** $y = -\frac{x}{2} + \frac{1}{2}$ **e** $(5, -2)$

SKILLS CHECK 2C (page 38)

1 a $(x - 3)^2 + (y + 2)^2 = 16$ **b** $(x + 5)^2 + y^2 = 25$

2 a $(2, 4), 6$ **b** $(0, -3), 4$

3 a $x^2 + y^2 + 2x - 4y - 20 = 0$ **b** $x^2 + y^2 + 6x - 8y = 0$

   c $x^2 + y^2 - 4x - 12y + 20 = 0$ **d** $x^2 + y^2 + 4x - 4y - 8 = 0$

4 a $(x - 1)^2 + (y - 2)^2 = 5^2, (1, 2), 5$

   b $(x + 5)^2 + (y + 12)^2 = 13^2, (-5, -12), 13$

   c $(x - 3)^2 + (y + 5)^2 = 4^2, (3, -5), 4$

   d $(x + 1)^2 + (y - 3)^2 = 7, (-1, 3), \sqrt{7}$

5 $(x - 2)^2 + (y + 5)^2 = 5^2$

6 $(x - 1)^2 + (y + 4)^2 = 10$

7 $(1, 1), \sqrt{5}; (7, 4), 2\sqrt{5}$

8 a $-\dfrac{2}{3}$ **b** $y = \frac{3}{2}x - \frac{11}{2}$

9 a $(x - 2)^2 + (y + 3)^2 = 4^2$ **b** $4, (2, -3)$ **c** $\begin{pmatrix} 2 \\ -3 \end{pmatrix}$

10 $x^2 + y^2 - 10x + 6y - 15 = 0$

SKILLS CHECK 2D (page 43)

1 c $(0, 4), \sqrt{17}$ **d** $x^2 + (y - 4)^2 = 17$

2 $k = 4, -1$

3 b $x = 8$ **c** $p = 8, q = -1$

4 c Yes. C on perp. bisector of AB

5 a $(1, 2)$

6 a $y = 2x - 2$ **b** $y = -x + 4$ **c** $(2, 2)$

7 $y = \frac{1}{3}x - \frac{10}{3}$

SKILLS CHECK 2E (page 45)

1 a $y = -x + 4$

2 $y = \frac{1}{2}x + \frac{3}{2}$

3 $y = -\frac{1}{3}x + \frac{8}{3}$

4 $y = 4x - 3, y = -\frac{1}{4}x - 3; (0, -3)$

5 $y = \frac{1}{2}x + 2, y = 2x - 6$

6 a $(2, -5), 5$ **b** $3x - 4y - 26 = 0$

SKILLS CHECK 2F (page 47)

1 a $(3, 4), (4, 3)$ **b** $(2, 4)$ **c** $(1, 3), (4, 6)$ **d** $(-2, -1)$

2 $x = 1, y = -2$

4 a $x = 1, y = 0$ **b** $x = 0, y = 0; x = 6, y = 0$

   c $x = 2, y = 8$ **d** None

5 $k = 1$

6 b $(-3, 4), (-4, 3)$

7 a $2x + 3y = 17$ **b** $P(1, 5)$ **c** $(x - 4)^2 + (y - 3)^2 = 13$

Exam practice 2 (page 48)

1 a $\dfrac{3}{2}, y = \frac{3}{2}x$

2 a $C(5, 1), \sqrt{29}; (x - 5)^2 + (y - 1)^2 = 29$

3 a $a = -3, b = 6$ **b** $\left(-\dfrac{3}{2}, 3 \right)$ **c** $y = -\frac{1}{2}x + \frac{9}{4}$

4 a i $k = 4$ **ii** $\dfrac{5}{2}$ **b** $2x + 5y + 3 = 0$ **c** $p = 5$

5 a $7, (-2, 7)$

6 b $y = 2x - 5$ **c i** $x - 3y + 15 = 0$ **ii** $(6, 7)$

7 a $\dfrac{5}{3}$ **b ii** $3x + 5y = 2$ **c** $(7, 3)$

8 a $7x + 6y = 61$ **b** 17   **c** $6x - 7y = 11$

9 a $x + 7y = 50$ **c ii** $(7.5, 2.5)$ **iii** $\frac{5}{2}\sqrt{10}$

10 a $y = 5 - x$ **b** $y = x - 1$ **c** $(3, 2)$ **d** $(x - 5)^2 + (y - 4)^2 = 8$

11 a i -3 **ii** $3x + y = 4$ **b i** $(2, -2)$ **ii** $\frac{1}{3}$ **iii** $p = -1$

12 a $(1, -3), 5$ **b** $5\sqrt{2}$

13 a $x = 3, y = 2$ **b** line is tangent to curve

14 b $P(-1 - \sqrt{7}, 1 + \sqrt{7}), Q(-1 + \sqrt{7}, 1 + \sqrt{7})$

15 a i $(0, 5)$ **ii** 3   **c** $\left(\dfrac{12}{5}, \dfrac{16}{5} \right)$, L is a tangent to C.

SKILLS CHECK 3A (page 52)

1 a $4x^3$ b $21x^2 - 4x$ c $32x + 8$ d $2x - \frac{1}{2}$

2 a $3x^2 - 4x + 1$ b $6x^3 + \frac{5}{2}$ c 0

3 56
4 $-3\frac{1}{2}$
5 b -12
6 a $2x + 5$ b $(-1, -8)$
7 $8x + 17$
8 $3x^2 - 4$
9 $5 - 6y$
10 a $3x^2 - 6x$ b $(1, -2)$

SKILLS CHECK 3B (page 55)

1 $y = 3x - 5$
2 a 2   b 5   c $y = 5x - 3$ d $x + 5y - 11 = 0$
3 a i $x > 2$ ii $x < 2$
   b i $x > 0$ ii $x < 0$
   c i $x > 1, x < -1$ ii $-1 < x < 1$
4 a -2 b $y = \frac{1}{2}x + 3$ c $\frac{5}{2}$
5 a $y = \frac{1}{2}x + \frac{9}{2}$ b $\left(\frac{3}{2}, \frac{21}{4}\right)$
6 a $y = -6x$ b $(1, -2)$
7 a $x > \frac{1}{12}$
8 a $y = 4x - 12$ b $A(3, 0), B(0, -12)$ c 18

SKILLS CHECK 3C (page 60)

1 a $(0, 0)$ min, $\left(-\frac{2}{3}, \frac{4}{27}\right)$ max   b $(1, -2)$ min, $(-1, 2)$ max
   c $(-2, 26)$ max, $(1, -1)$ min
2 $x = -\frac{1}{2}$ max, $x = \frac{2}{3}$ min
3 500,000 m²
4 10,000
5 b 1, 3   c 16
6 c $360x - 144x^2$; 2.5   d $A = 360x - 84x^2$

Exam practice 3 (page 61)

1 a $6x(x + 1)$ b $(0, 1)$ min, $(-1, 2)$ max   c $x > 0, x < -1$
2 a $\frac{1}{2}$ b i $6x^2 - 2x + 8$
3 a $4x(x^2 - 6x + 8)$ b $x = 0, 2, 4$ c 8:24 am
4 c 1   d 18 cm³
5 a $V = hx^2, A = x^2 + 4xh$ b i $10\sqrt{10}$ iii $5000\sqrt{10}$
6 a -1 b 1   c $y = x - 2$ d 3

7 $A\left(\frac{5}{2}, -\frac{17}{4}\right)$

SKILLS CHECK 4A (page 65)

1 a $\frac{4}{3}x^3 + c$ b $\frac{2}{15}x^3 + c$ c $-6x + c$

2 a $y = \frac{1}{6}x^6 + x^3 + c$ b $y = \frac{2}{5}x^5 - \frac{1}{8}x^4 + c$ c $y = \frac{2}{3}x^3 + \frac{7}{2}x^2 + c$

---

d $y = -\frac{3}{2}x^2 + 4x + c$ e $y = -\frac{1}{10}x^4 - x + c$ f $y = 2cx + k$

3 a $2x^2 - 5x - 12$ b $\frac{2}{3}x^3 - \frac{5}{2}x^2 - 12x + c$

4 a $A = \frac{1}{2}t^8 + c$ b 127.5

5 $y = \frac{5}{2}x^2 - 3x$

6 $y = x^3 + x^2 + 4x - 6$

7 $4x - \frac{1}{3}x^3 - 3\frac{2}{3}$

8 a $\frac{4}{3}x^3 + 6x^2 + 9x + c$ b $\frac{4}{3}x^3 - 9x + c$

SKILLS CHECK 4B (page 69)

1 a 7   b 6.4   c 1.25
2 a 12   b -5.25
3 a $\frac{1}{3}$ b $10\frac{2}{3}$
4 16 square units
5 a $A(-1, 0), B(2, 0)$ b 4.5 square units
6 b $\frac{1}{3}$ square units
7 a $x = 1, y = 3; x = -2, y = 6$
   b i $P(-2, 6), Q(1, 3)$ ii $4\frac{1}{2}$ square units

Exam practice 4 (page 70)

1 $y = x^3 - \frac{1}{2}x^2 - 5$
2 a $12x^2$ b i $x^4 + x + c$ ii 16 square units
4 a 6, $6x - y - 6 = 0$ b $\frac{1}{4}$ square units
5 b $C(-2, 0)$
6 a i $1 - 8x^3$ iii $\frac{3}{8}$ b i $\frac{1}{2}x^2 - \frac{2}{5}x^5 + c$ ii $\frac{9}{80}$
7 a $P(-\sqrt{3}, 5 - 2\sqrt{3}), Q(\sqrt{3}, 5 + 2\sqrt{3})$ b $4\sqrt{3}$
8 a 16   b i -12 ii $Q(0, 24)$ c 8
9 a $f(3) = -2, f(4) = 0$ b $(x - 1)(x - 2)(x - 4)$
   c i $3x^2 - 14x + 14$ ii Decreasing, $f'(3) = -1$
      iii $\frac{1}{4}x^4 - \frac{7}{3}x^3 + 7x^2 - 8x + c$ iv $\frac{5}{12}$

Practice exam paper (page 73)

1 a $-4 + 14\sqrt{2}$ b $9 - 3\sqrt{2}$
2 a $(2, 1)$ b i $y = \frac{1}{2}x$
3 a $x > 3.5$ b $0 < k < 8$
4 b i $\frac{x^5}{5} - \frac{2x^3}{3} + x + c$ ii $\frac{8}{15}$
5 c 0, $f(x) = (x + 1)(x - 2)(x + 4)$
6 a translation by $\begin{pmatrix} -2 \\ -4 \end{pmatrix}$
   b i $x = 0, -4$ ii $(-4, 0), (0, 0)$ iii $(-2, -4)$
   c ii $(-3, -3), (1, 5)$
7 a $36 - 6x - 12x^2$ b $x = -2, \frac{3}{2}$
   c i $-6 - 24x$ ii 42, -42 d $(-2, 20)$
8 a $(x - 1)^2 + (y + 2)^2 = 100$
   b $C(1, -2)$ c i $3y = 4x - 10$

SINGLE USER LICENCE AGREEMENT FOR CORE 1 FOR AQA CD-ROM

IMPORTANT: READ CAREFULLY

WARNING: BY OPENING THE PACKAGE YOU AGREE TO BE BOUND BY THE TERMS OF THE LICENCE AGREEMENT BELOW.

This is a legally binding agreement between You (the user or purchaser) and Pearson Education Limited. By retaining this licence, any software media or accompanying written materials or carrying out any of the permitted activities You agree to be bound by the terms of the licence agreement below.

If You do not agree to these terms then promptly return the entire publication (this licence and all software, written materials, packaging and any other components received with it) with Your sales receipt to Your supplier for a full refund.

YOU ARE PERMITTED TO:

- Use (load into temporary memory or permanent storage) a single copy of the software on only one computer at a time. If this computer is linked to a network then the software may only be used in a manner such that it is not accessible to other machines on the network.

- Transfer the software from one computer to another provided that you only use it on one computer at a time.

- Print a single copy of any PDF file from the CD-ROM for the sole use of the user.

YOU MAY NOT:

- Rent or lease the software or any part of the publication.

- Copy any part of the documentation, except where specifically indicated otherwise.

- Make copies of the software, other than for backup purposes.

- Reverse engineer, decompile or disassemble the software.

- Use the software on more than one computer at a time.

- Install the software on any networked computer in a way that could allow access to it from more than one machine on the network.

- Use the software in any way not specified above without the prior written consent of Pearson Education Limited.

- Print off multiple copies of any PDF file.

This licer   he software

PEARSON EDUCATION LIMITED RESERVES   S LICENCE BY WRITTEN NOTICE AND TO TAKE
ACTION TO RECOVER ANY DAMAGES SUFF   I LIMITED IF YOU BREACH ANY PROVISION OF

Pearson Education Limited and/or its licensors own the software.
You only own the disk on which the software is supplied.

Pearson Education Limited warrants that the diskette or CD-ROM on which the software is supplied is free from defects in materials and workmanship under normal use for ninety (90) days from the date You receive it. This warranty is limited to You and is not transferable. Pearson Education Limited does not warrant that the functions of the software meet Your requirements or that the media is compatible with any computer system on which it is used or that the operation of the software will be unlimited or error free.

You assume responsibility for selecting the software to achieve Your intended results and for the installation of, the use of and the results obtained from the software. The entire liability of Pearson Education Limited and its suppliers and your only remedy shall be replacement free of charge of the components that do not meet this warranty.

This limited warranty is void if any damage has resulted from accident, abuse, misapplication, service or modification by someone other than Pearson Education Limited. In no event shall Pearson Education Limited or its suppliers be liable for any damages whatsoever arising out of installation of the software, even if advised of the possibility of such damages. Pearson Education Limited will not be liable for any loss or damage of any nature suffered by any party as a result of reliance upon or reproduction of or any errors in the content of the publication.

Pearson Education Limited does not limit its liability for death or personal injury caused by its negligence.

This licence agreement shall be governed by and interpreted and construed in accordance with English law.